A Trip Through Death Valley's Geologic Past

(The Magnificent Rocks of Death Valley)

Kenneth E. Lengner

story of the
geological
ples, dating
it. Products
in lieu of a
discussion of the geological processes. Numerous photographs of rock samples and historical photos are used to delve into the past. This is not intended to be a definitive, scholarly geological work. Rather, it is meant to be an enlightening and entertaining trip through Death Valley's geologic past for Death Valley and geology enthusiasts. The trip is only a step towards, "What is it and where did it come from?"

ISBN: 978-0-9820883-2-6

First Printing 2009
Library of Congress
Cataloging-in-Publication Data
Lengner, Kenneth
A Trip Through Death Valley's Geologic Past
Kenneth E. Lengner
Includes bibliographical references
TM Deep Enough Press

Billion year old diamictite in Sperry Wash contains rocks from even earlier times.

Dedication

Bennie W. Troxel, Trailblazer and Geologist

In addition to the approximately 130 professional papers he authored or coauthored, field trips he led, graduate students he mentored, dinners he helped to cook for visiting geology students and faculty, and patient efforts trying to teach some of us about the region's geology; Bennie found the time to conceptualize and collect samples for his rock trail now seen in front of the Museum in Shoshone, California. He scoured the desert of the Death Valley region for over 55 years, from north to south, from east to west, and from the bottom of the basins to the top of the mountains.

Bennie W. Troxel

Bennie was born in Osawatomie, Kansas in 1920. He moved to Los Angeles, California in 1939 and subsequently got a job at Northrop Aircraft in Hawthorne California where he assembled aircraft. During World War II, he served in the U. S. Army Air Corps in such far away places as India and China.

Following service in the U. S. Army Air Corps during World War II, he briefly returned to work at Northrop Aircraft in Hawthorne before enrolling in Compton Junior College where a few introductory courses in geology from instructors Robert E. Stevenson and Gordon Oakeshott peeked his interest. He enrolled at UCLA where he studied under W. C. Putnam, J. Murdock, Don Carlisle, John Crowell, Clem Nelson Cordell Durrell, U. S. Grant IV, Jim Gilluly, "Parky" Parkinson, and others. He eventually got his Masters Degree in geology from UCLA.

Bennie then went to work at the California Division of Mines where he met Dick Jahns and Bennie's fellow Death Valley geologist, Lauren Wright. (The two were recently honored by a book entitled *"Fifty Years of Death Valley Research."*). Soon thereafter (~1952) he started the geological mapping of the Avawatz Mountains where he lived in a cave with a mouse and a bat. Bennie and Lauren met famed geologist, Levi Noble, who convinced them to work on a project in the Virgin Springs area. They maintained a long, professional association with Levi through which they met some of Levi's peers such as Chester Longwell, Bill Pecora, Charles Anderson, and Charles Denny. Besides mapping the Avawatz, Bennie moved on to mapping and reconnaissance in the Funeral Mountains, parts of the Shoshone and Tecopa area, the Greenwater Range, Saddle Peak Hills, Last Chance Range, Slate Range, Spring Mountains, Fort Irwin, etc.

Lauren and Bennie's joint work in the Funeral Mountains and Virgin Spring area of the Black Mountains led them to the recognition of extension tectonics in Death Valley, a stunning achievement in the geology world.

Early in his career, he became intrigued with the colorful and somewhat enigmatic Kingston Peak Formation (KP) which had been first described by one of his heroes, Foster Hewett. Some believe that the KP's origin is glacial deposits. Bennie's analysis indicates that "dropstones" in the KP matrix were derived from melting icebergs far from the "dropstones" point of origin. He prefers to say that the stones are lag features from deep-water fans.

Interest in Death Valley geology grew and associations with other geologists (e.g. Charles Hunt, Don Mabay, Donald Curry, Mitchell Reynolds, and Jack Stewart) formed. Bennie spent two summers teaching field geology for the University of Nevada Las Vegas. Students and instructors from universities throughout the U. S. and Europe began arriving through which he met instructors and researchers Terry Pavlis, Laura Serpa, Brian Wernicki, Jim Calzia, Matt McMackin, Marli Miller, and a host of others.

A favorite comment is, " In Death Valley one is apt to be confused by a huge abundance of data."

In a recent book honoring Bennie and Lauren, Bennie's last comment was, "Lastly, I am forever grateful to my beloved wife Betty and our two children who have been extremely tolerant of my many absences from home while working in Death Valley.

Those who know Bennie consider him a consummate geologist and a sterling human being.

Ken Lengner
Pahrump, Nevada
October 2009

About the Author

Ken Lengner, originally from Wanaque, New Jersey, was an Aerospace Engineer for 34-years. He started his career as a Systems Engineer on the Apollo Moon Landing Program and retired from Boeing as Chief Project Engineer of Space Shuttle-Orbiter Subsystem Managers. He has been a full time resident of the Death Valley region for the past ten years. For approximately 30-years, a parallel "career" entailed frequent visits to Death Valley National Park and surroundings. He is an avid historian, photographer and student of the Death Valley region's geology, biology, paleontology, and paleogeography. He does volunteer work for Death Valley National Park and guides tours in the area. He has authored and published four other books on the Death Valley region, edited yet another, and has more in progress.

Horn coral, Pennsylvanian, 310 MY, Bird Spring Fm., southern Death Valley.

Acknowledgements

Bennie W. Troxel is the renowned geologist who, with permission of Death Valley National Park and Bureau of Land Management, collected large sized samples of Death Valley region rocks and arranged them in chronological order at the Shoshone Museum. His objective was to provide an educational trail through Death Valley's geological history. That rock trail provided inspiration for this book. Through the past twelve years Bennie has provided patient instruction to this novice geologist as well as a lasting friendship. He has helped edit this document and answered many questions during its writing.

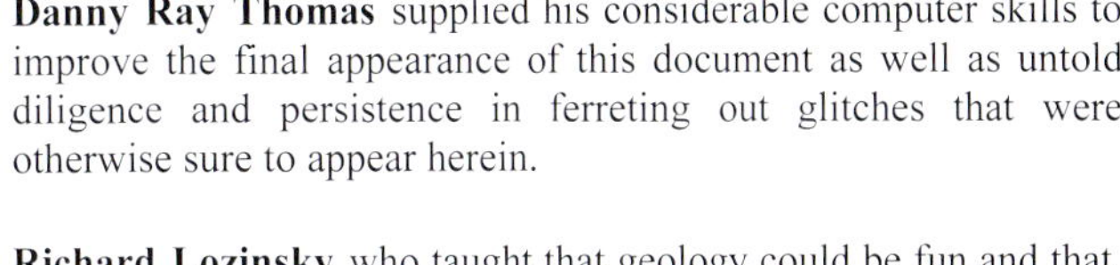

Danny Ray Thomas supplied his considerable computer skills to improve the final appearance of this document as well as untold diligence and persistence in ferreting out glitches that were otherwise sure to appear herein.

Richard Lozinsky who taught that geology could be fun and that, "Rocks do not lie."

Jim Calzia, Marli Miller, Lauren Wright, Terry Pavlis, and a host of other professional geologists who took the time to explain geological phenomenon and let me listen to their technical musings.

Blair Davenport, Curator Death Valley National Park, and Curatorial Staff **Emily Pronovost** and **Ann Powell** for helping with innumerable research projects.

Death Valley National Park supplied historical photographs.

All photographs were supplied by the author unless otherwise noted.

Gastropod and receptaculitid fossils in the Pogonip Group, Ordovician, ~400 MY, eastern Death Valley region.

Table of Contents

Oldest rocks in the Death Valley region are metamorphic gneiss of the crystalline basement dated at 1.7 billion years old.

Introduction

Over 20 years ago, renowned Death Valley geologist, Bennie W. Troxel, got the idea to collect large-sized samples of rocks that typify and tell the geologic story of the Death Valley region. He collected the earliest Death Valley region's rocks from when the region was on a super continent known as Columbia, rocks from when Columbia was torn asunder, rocks from when the region was beneath a shallow sea, rocks from the Age of Dinosaurs when the region had emerged from that sea, rocks from when the region was a flat plain on which giant Titanotheres roamed, and rocks from the time during which the region was once again torn apart. He arranged these rocks in the sequence in which they had been deposited in order to be able to describe Death Valley's evolution… a "trail though time."

This trail is located in Shoshone, California in front of the Shoshone museum. It consists of approximately 100 rocks weighing from 10 to 50 pounds, some of which are on wood pedestals with placards describing the associated rocks. The rocks document the geologic history by providing characteristic samples chronologically ranging from the earliest Death Valley rocks (metamorphic gneiss) to relatively contemporary rocks such as borates and fanglomerates. As one progresses through time while traveling along the trail, rocks such as augen and foliated gneiss, conglomerate, cross-bedded sandstone, purple shale, diabase, talc, chert and dolomite, diamictite, dolomite with vent tubes, rocks with ripple marks and oolites, quartzite, limestone and more dolomite, rhyolite to basalt, granite to gabbro, borate, and fanglomerate are seen. The journey entails Death Valley's crystalline basement; Crystal Springs, Beck Springs and Kingston Peak Formations; Noonday Dolomite; Johnnie Formation; Stirling Quartzite; Wood Canyon Formation; Zabriskie Quartzite; Carrara and Bonanza King Formations; Eureka Quartzite; Tin Mountain Limestone; Titus Canyon Formation; volcanic and plutonic rocks; and evaporates.

An early photo of the Shoshone Museum rock trail under construction.

Those of us who accompanied Bennie on field trips and rock collecting expeditions learned much about the region because he always spoke in such a manner that whoever he was speaking to could understand him. Those lucky enough to be with him when a suitable sample was found, got to haul it back to the truck and then to the Shoshone Museum. After all, there were certain privileges associated with being the Senior Geologist on that trip. Sampling inside Death Valley National Park was always done with permission.

This book was inspired by Bennie W. Troxel's many studies in the Death Valley region and utilizes some of the rocks along his "trail through time" and other rocks as guideposts. Photographs of specific rocks along Bennie's trail are denoted by **T**. Other photos and illustrations found elsewhere do not include **T**.

The main objective of this document is to provide an overview of the Death Valley region's geologic history or a virtual "trip through Death Valley's geologic past." This objective will be accomplished by first providing a few fundamentals such as the basics of rock types, geologic time, major events in the long geologic history of Death Valley region, and a pictorial summary of locations of rock types. Next, chronologically, rock samples are used as guideposts or slices of time for brief discussions concerning the rocks' compositions, where they currently can be found, how they were originally deposited, and what Death Valley environment (i.e. climate, plants, and animals) was like at the time of deposition. This "trip" through time is designed to provide a top level view of the geologic history of the Death Valley region for non-professional geologists. Next, a collection of photographs of Death Valley rocks is included to further the reader's appreciation of the region. A stratigraphic column is provided in the Appendix to facilitate a further understanding of the chronology and complexity of the region's geology.

T **This symbol is used throughout text to identify rock samples found along Troxel trail, Shoshone Museum.**

The sun rises on a receding Amargosa River as flooding continues to sweep sediments downstream into the Death Valley basin where new rocks are being formed as others weather and erode. The receding waters created ripple marks like those solidified in some of the Death Valley region's oldest rocks.

"The present is the key to the past." *James Hutton/Charles Lyell*

Rocks and the Rock Cycle

Because we are going to progress through time looking at Death Valley **rocks** and then look at the "magnificent **rocks** of Death Valley," it is appropriate to have a discussion of, "What is a rock?" Too simple? Read on. Essentially, rocks are naturally formed, consolidated material composed of grains of one or more minerals. There are three types of rocks. **Igneous** rocks are formed from cooling magma. When magma is ejected above the surface by volcanoes, it cools relatively rapidly and forms **extrusive igneous** rocks. When plutons rise from the mantle towards the crust and magma cools slowly beneath the surface, the igneous rock has distinct crystals and is said to be **intrusive igneous** rocks. **Sedimentary** rocks are formed from lithification of sediments, precipitation from solution, and consolidation from plant and animal matter. Sediments are derived from weathering of igneous, sedimentary, or the third type of rock… metamorphic rock. **Metamorphic** rocks are formed from pre-existing rocks. They change form (without melting) due to pressure from layers of overburden, temperature from sources such as upwelling magma in their proximity, and tectonic forces creating compression.

The rock cycle in **Figure 1** addresses the three types of rocks, touches on how they form, and indicates how one type of rock is formed by another. For example, weathering of all three types of rocks creates sediments which can become a new sedimentary rock via compaction. That new sedimentary rock can be changed to a metamorphic rock by heat from rising plutons. If that new metamorphic rock is near a converging tectonic plate boundary, it may be pulled down into the magma. The magma may rise in a pluton, cool beneath the surface, and form the intrusive igneous rock known as granite. Rocks aren't destroyed, they are cycled. An introduction to rocks we shall meet on the trail is provided in **Table 1**.

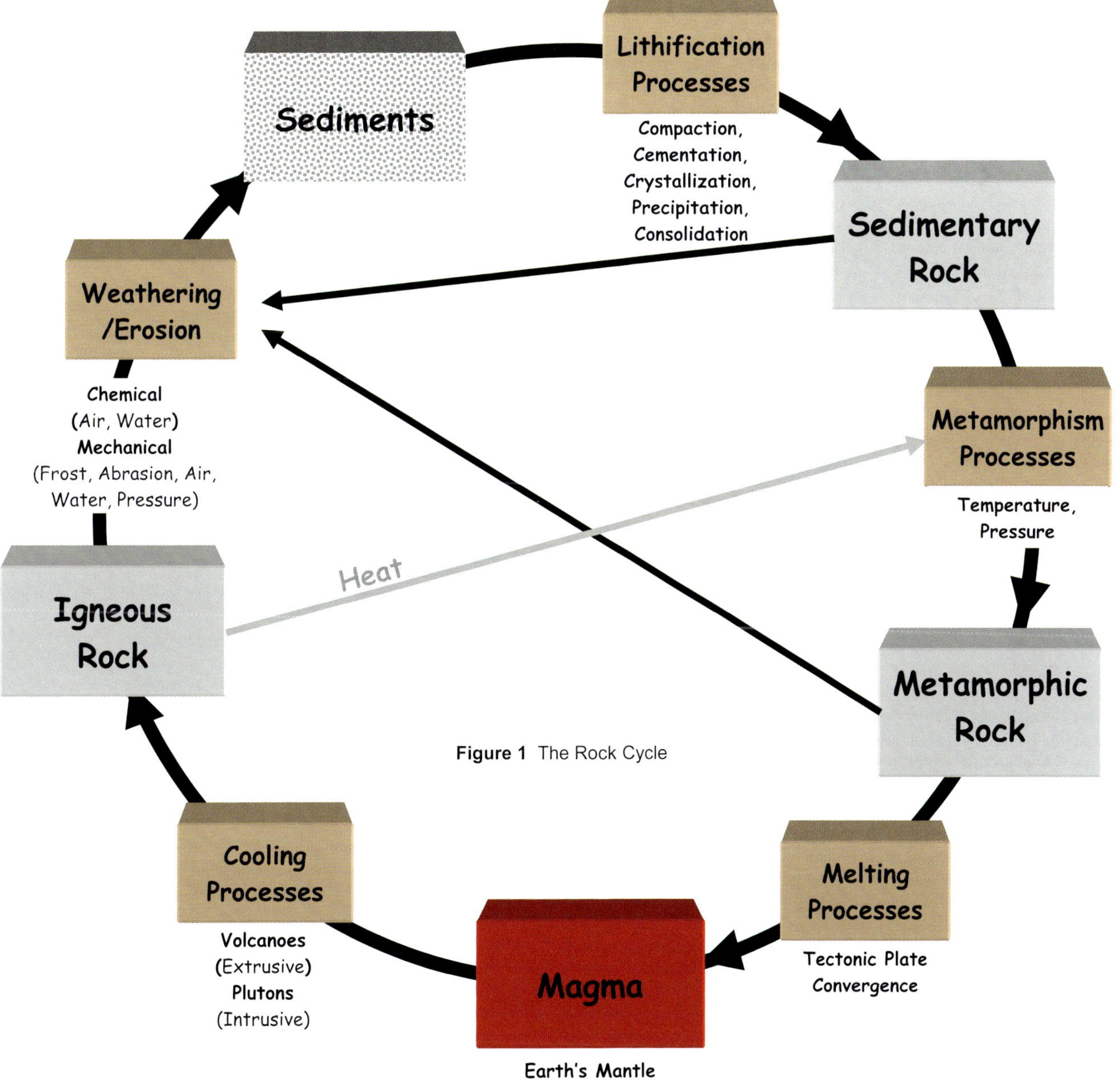

Figure 1 The Rock Cycle

Table 1 Introduction to Some Death Valley Region Rocks

Intrusive Igneous (pluton)	Extrusive Igneous (volcano)	Sedimentary	Metamorphic
Granite-intrusive counterpart to rhyolite; coarse-uneven-grained; predominately feldspar and quartz (silica). White, gray.	**Rhyolite**-extrusive counterpart to granite; fine grained to porphyritic (many visible grains of different size scattered in dense matrix). Tan, beige. (High silica, violent volcano).	**Conglomerate**- Different lithology round cobbles, fine matrix. / **Breccia**- Different lithology angular cobbles, fine matrix.	**Gneiss**- *from* intrusive igneous and sedimentary (e.g. shale, sandstone, conglomerates) rocks; medium to coarse-grained; parallel, folded bands.
Granodioite-coarse-uneven-grained; 2/3 light colored minerals (quartz, feldspars).	**Dacite**-fine grained to porphyritic (many visible grains of different size scattered in dense matrix).	**Mudstone**-dense with unknown proportions of silt and clay sized particles; massive and thick layers. Light to dark gray. Knife scratches. **Siltstone** a massive mudstone in which silt predominates over clay; hard thin layers.	**Schist**- *from* extrusive igneous and sedimentary (e.g. shale, sandstone, conglomerates) medium to coarse-grained; parallel mineral orientation; sparkling luster.
Diorite-intrusive counterpart to andesite; coarse-uneven-grained with more plagioclase feldspar than iron and magnesium minerals. Light and dark minerals in equal amounts.	**Andesite**-extrusive counterpart to diorite; fine-grained. Pink.	**Shale**-dense with silt or clay sized particles; thin-bedded; harder than mudstone. Light to dark gray. Knife scratches. From quiet environment of lake or ocean bottom.	**Slate**- *from* shale; fine-grained, splits to thin wavy sheets.
Gabbro-intrusive counterpart to basalt; coarse-uneven grained; iron and magnesium minerals and gray, plagioclase feldspar. Black, Green. At least 2/3 dark minerals.	**Basalt**-extrusive counterpart to gabbro; fine-grained to porphyritic, vesicular; with iron and magnesium minerals and gray, plagioclase feldspar. Dark to black. ("benign" volcano).	**Sandston**-even sized grains generally rounded; thick-bedded, gritty texture. Three types: quartz (>90% of grains is quartz), arkose (>25% feldspar), graywache (>15% is fine grained matrix).	**Quartzite**- *from* quartz rich sandstone; white to dark, thick-bedded, crystalline texture.
Hornblendite-predominately hornblende which is a rock forming mineral found in igneous and metamorphic rocks. Mostly iron and magnesium Dark green/ black.	**Tuff**-layers in varying thickness; mostly light colored; gritty; possible angular rock or mineral fragments. Ash falls or ash flows.	**Limestone**-fine-grained; crystalline texture; mostly single mineral (calcium carbonate). Precipitate, marine shells. Effervesces in weak hydrochloric acid. $CaCO_3$.	**Marble**- *from* limestone; even granular; lustrous, white to dark gray.
Diabase-a variety of gabbro. Black, Black-green.	**Lava**-rhyolite, dacite, trachyte, andesite, basalt flowing from volcano. Dense to vesicular (porous, light weight).	**Dolomite**-Limestone that has some of the calcium element replaced by magnesium. $CaMg(CO_3)_2$.	**Dolomitic marble**- *from* dolomite.
		Calcite-rhombohedral; $CaCO_3$, many colors. / **Travertine**- Hot springs; color banded, $CaCO_3$.	**Hornfels**- *from* shale or basalt ; fine-grained, dark, scratch glass, some coarser minerals, conchoidal fracture.
		Evaporites-from drying lakebeds: Borates (contain element boron); Gypsum ($CaSO_4 \bullet 2H_2O$); Salts (NaCl, KCl).	**Talc**- *from* dolomite or ferromagnesian rock e.g. hornblendite. White, green; greasy.
		Chert-Silica oxide; lenses, nodules, or thin beds. SiO_2.	**Wollastonite**- *from* impure limestone. White, gray, translucent; fluorescent.

Overview of the Death Valley Region Geology

Geologic Time

To understand how the Death Valley region came to be, we first must get a grasp of "geologic time." **Table 2** defines the eons, eras, periods, and epochs that geologic time is divided into. The first thing to note is that geologic time encompasses a long time… a very, very long time. It begins when Earth's crust was formed 4.6 billion years ago and continues to the present. Mankind's reign on Earth is just a blink of geologic time. The second thing to note in the table is how it is divided into eons, eras, periods, and epochs and their associated time-frames. Scientists have divided up geologic time based on significant events. For example, the beginning of the Cambrian was originally defined at a time when scientists thought there was an explosion of life in Earth's oceans. In following sections, we will refer to these eons, eras, periods, and timeframes. Because the concept and spans associated with geologic time can be daunting, this table and text maintains a consistent color coding throughout to help orient the reader as to what timeframe he is in.

Table 2 Simplified Geologic Time Scale

Eon	Era	Period	Epoch
Phanerozoic Eon (0-570 MY)	Cenozoic Era (0-66 MY)	Quaternary Period (0-2.6 MY)	Holocene Epoch (0.0-10 KY)
			Pleistocene Epoch (.01-2.6 MY)
		Tertiary Period (2.6-66 MY)	Pliocene Epoch (2.6-5 MY)
			Miocene Epoch (5-23 MY)
			Oligocene Epoch (23-34 My)
			Eocene Epoch (34-56 MY)
			Paleocene Epoch (56-66 MY)
	Mesozoic Era (66-251 MY)	Cretaceous Period (66-146 MY)	
		Jurassic Period (146-202 MY)	
		Triassic Period (202-251 MY)	
	Paleozoic Era (251-542 MY)	Permian Period (251-299 MY)	
		Carboniferous Periods: Pennsylvanian (299-318 MY) Mississippian (318-359 MY)	
		Devonian Period (359-416 MY)	
		Silurian Period (416-444 MY)	
		Ordovician Period(444-488 MY)	
		Cambrian Period (488-542 MY)	
Proterozoic Eon (542-2500 MY)			
Archean Eon (2.5-3.85 BY) AND Hadean Eon (3.85-4.6 BY)			

1) BY= Billion years (ago), MY= Million years (ago), KY= Thousand years (ago)
2) 2009 Geologic time scale, rounded to integer. The exact numerical values defining the boundaries of the eons, eras, periods, and epochs is constantly being changed as more data arises.

Plate Tectonics Creates Death Valley Geologic Environments

Earth's surface is comprised of constantly, slow-moving, oceanic and continental tectonic plates which converge, diverge, or grind along each other. This dance led Death Valley into northern, southern, and equatorial latitudes. Essentially, plate tectonics created the climatic and geological environments that, in turn, created the Death Valley region.

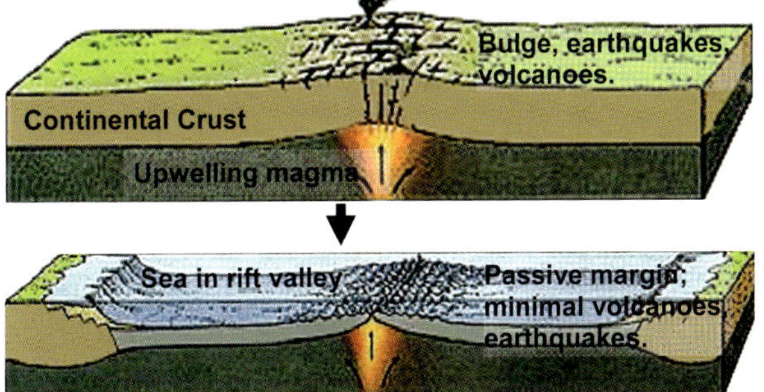

Figure 2 Possibly how Death Valley's Late Proterozoic to Late Paleozoic geologic environment formed. Magma (basalt) wells upward, rifts and continues to spread continental crust, and creates volcanoes and earthquakes. Rift valley formed, sea flooded in, basaltic oceanic plates formed and continued to diverge, **passive continental margin** formed. *From USGS*

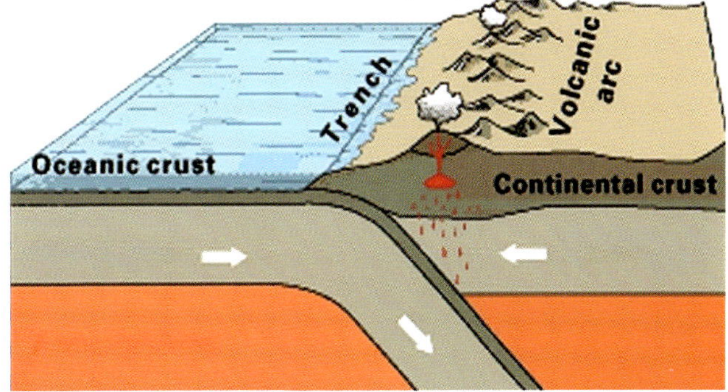

Figure 3 From the Late Paleozoic to Middle Cenozoic, the Death Valley region's geologic environment was driven by "subduction" where an oceanic crustal plate dove under the continental crust. An **active continental margin** resulted. The Death Valley region saw volcanoes, earthquakes, and mountain building. *From USGS*

Geologic Time and Death Valley Major Events

Earth and the Death Valley region have a long and complicated history. **Figure 4** has been provided to enhance the reader's understanding of geologic time and to identify major events. Note that this figure has a long line that winds back upon itself. That line's length attempts to represent the 4.6 billion years of geologic time while the very top, dot is modern man's presence. Major events occurring on Earth and in Death Valley are identified. For example, the oldest rock in Death Valley was deposited at a time (estimated 2.5 billion years ago) when there were no plants or animals on land, no animals lived in the oceans, and cyanobacteria was producing oxygen for us to breath. Later Death Valley's oldest rock was metamorphosed (dated 1.7 billion years ago in the Proterozoic Eon). Subsequently, the Death Valley region was primarily beneath a shallow sea for approximately a billion years until its final emergence at the beginning of the age of dinosaurs. **Table 3** provides you with another view of major Earth and Death Valley events.

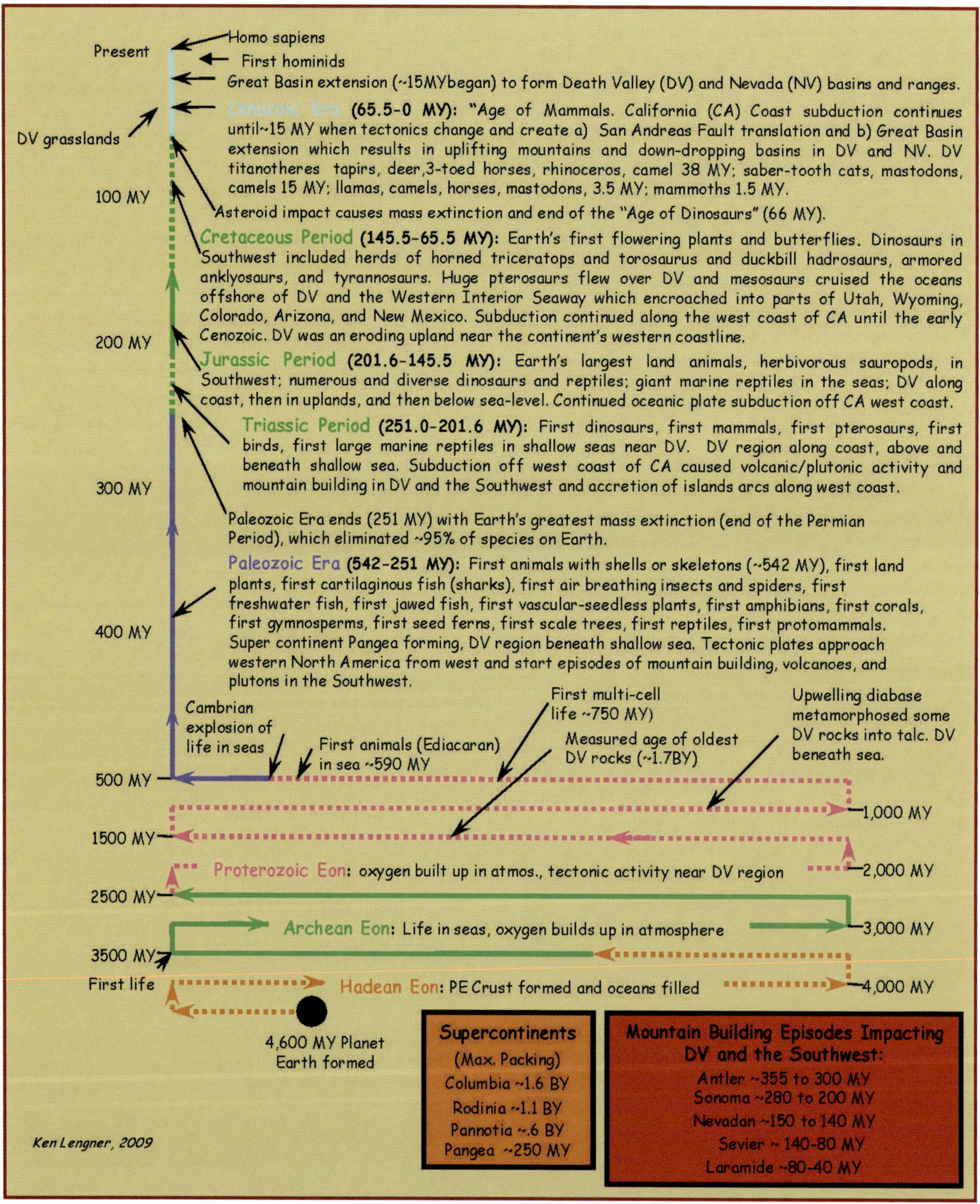

Figure 4 Death Valley Region Major Events Along a Time Scale

Geologic Time and Death Valley Major Events (continued)

Table 3 Earth and Death Valley's Major Events		
Time (MY) 2009 GSA	**Earth's Major Events**	**Death Valley's (DV) Major Events**
Holocene 0.0-0.01 "Age of Man"	End of the previous "Ice Ages." Giant mammals disappeared. *Homo sapiens* became dominant species.	Man arrived. Climate varied: wet and finally dry. Large mammal extinction. Sparse vegetation. <2-inch rain/yr. First DV mining ~ 1850.
Pleistocene .01-2.6 "Ice Age (s)"	Single hominid (*Homo sapiens*) by ~25 KY. Giant mammals. Alternating glacial and warm periods. Hominids make tools and migrate to Europe/Asia.	Lakes Manly (300 ft deep), Saline, Tecopa. Wetter and cooler than today. Plants 3,500 ft lower than today. Desert plants begin to evolve. Mammoths, mastodons, various camels, horses, canines, llamas.
Pliocene 2.6-5.3 "Recent"	Cooling and drop in sea levels began. Coexisting hominids evolve in Africa, *Australopithecus afarensis* et al.	Great Basin extension. DV mountain ranges rose and basins dropped. Lakes began to form. Mammoths, mastodons, camels, horses.
Miocene 5.3-23.0 "Moderately Recent"	First kelp forests and sea otters. First apes in Africa ~23 MY. Warmer than Oligocene and Pliocene.	Mountains rise/basins drop ("extension," ~15 MY) to begin forming today's topography. Region more arid. Animal life hindered by DV and NV volcanic activity.
Oligocene 23.0-33.9 "But a Little Recent"	Glaciers. Lower temperatures shrank forests and grasslands spread. Modern herbivore types (horses) appeared.	DV was flat savannahs with streams and occasional volcanic activity. Titanotheres, 3-toed horses, camels, and rhinoceros.
Eocene 33.9-55.8 "Dawn of the Recent"	Odd and even toed herbivores. High temperatures began to decrease. Savannahs. First primates.	Mountains eroded. During late Late Eocene, plates west of DV began transition from convergent to lateral boundary. Eventually lateral DV faults.
Paleocene 55.8-65.5 "Old Recent" Begin "Age of Mammals"	Age of mammals began. First flesh-eating placental mammal. Large carnivorous birds. Subtropical climate.	Active margin until Middle Miocene, tectonic plate motion from west. Thrust forces fold previously deposited DV rock layers. Volcanic lava and ash. Plutons rose beneath future Panamint, Slate, Coso, Argus, and Saline Ranges and Avawatz, Granite, and Cottonwood Mtns. DV permanently above ocean. Minimal sedimentary rocks formed. DV was eroding uplands. (No dinosaur fossils found in DV region.)
Cretaceous 65.5-145.5 "Age of Dinosaurs"	65.5 MY asteroid impact. Mass extinction of non-avian dinosaurs. First flowering plants.	
Jurassic 145.5-201.6 "Age of Cycads"	Largest land animals - sauropods, First birds. Pangea breakup.	
Triassic 201.6-251 "Age of Reptiles" also began "Age of Dinosaurs"	Pangea straddled equator and rifting. Large, carnivorous reptile: Dimetrodon. Abundant cycads. First mammals. First dinosaurs. Minor extinction.	Active Margin. Subduction zones formed as plates converged west of DV. DV above and below sea level (Butte Valley marine). Eventually became topographic high (minimal sedimentary rocks formed) and erosion occurred.
Permian 251—299 "Age of Amphibians"	Supercontinent Pangea formed. Bone headed proto-mammals. Earth's largest mass extinction ~251 MY.	Active margin begins in Pennsylvanian. During majority of Paleozoic, DV was on passive margin beneath shallow, rising and falling ocean. Carbonate rock layers deposited with marine fossils: trilobites, oolites, oncolites, and burrowing worms. Occasional terrestrial rock deposits. Paleozoic seas intermittently extended far inland.
Pennsylvanian 299-318 "Age of Ferns"	First reptiles laid eggs. Forests source of future coal. Pangea forming.	
Mississippian 318-359 "Age of Amphibians" Begin "Carboniferous"	First seed ferns and scale trees. Abundant amphibians. Phytoplankton, zooplankton, echinoderms in oceans.	
Devonian 359-416 "Age of Fishes"	First amphibians and first vascular seed-bearing plants. Insects present.	
Silurian 416-444 "Age of Fishes"	First air-breathing insects, jawed fish, freshwater fish, and vascular-seedless plants. Warmer climate. Trilobites.	
Ordovician 444-488 "Age of Marine Invertebrates"	First land plants (non-vascular) and first corals. Trilobites, brachiopods, gastropods, nautoloids, graptolites.	
Cambrian 488-542 "Age of Marine Invertebrates"	Explosion life in oceans. First skeletal and shelly life (trilobites). First carnivores.	
Proterozoic 542-2500 "First Life"	First animals (Edicara) ~.590-635 BY. First eukaryotes ~1.6 BY Oxygen content increased ~3.5-1.8.	DV submerged and region rifting ~1.1 BY. Oldest date DV rocks 1.7 BY-extrapolated to 2.5 BY. DV on continent of Columbia > 1.7 BY.
Archean 2.5-3.8 BY	~3.5 BY First sea life (prokaryotes, bacteria) added oxygen and formed stromatolites.	Death Valley Region - No rock record.
Hadean 3.8-4.6 BY	PE crust formed, oceans filled.	Death Valley Region - No rock record.

A Geological Summary and the Results

During an estimated 2.5 billion years of Death Valley region geologic history, many events transpired which created a wide diversity of rocks. It is a history of a region torn apart, submerged beneath the seas, crushed by onrushing tectonic plates, and a region currently being torn apart... yet again. Sea levels rose and fell, mountains were uplifted and eroded, climates were hotter and colder than today, plant and animal species came and went as a matter of cataclysmic mass extinctions or gradual adaptations to changing environments, and tectonic plates drifted into the southern and northern latitudes and continued their dance as ancient supercontinents formed and dismantled.

The earliest Death Valley rocks were sedimentary and deposited during the Early Proterozoic Eon at an extrapolated date of 2.5 BY on a speculative ancient continent known as Columbia. From approximately 2.5 to 1.7 BY, the region is believed to have been under the sea for a portion of the time. By 1.7 BY, overburden had metamorphosed the original rocks and the region had been intruded by granite. From before 1.7 to 1.1 BY, the region was primarily under a sea but may have briefly been on another speculative continent called Rodinia. At 1.1 BY, with Death Valley beneath the sea, rifting in the Death Valley region resulted in a diverging boundary with upwelling of hot magma that "cooked" some Middle Proterozoic Eon rocks and produced talc deposits that would later be mined. By approximately 600 MY, the Death Valley region had progressed far from the diverging boundary and was on a passive continental margin where marine rocks like limestone (calcium carbonate, $CaCO_3$) and dolomite (calcium magnesium carbonate, $CaMg(CO_3)_2$) continued forming. Death Valley was part of or near supercontinents such as Pannotia (600 MY) and Pangea (250 MY) as they formed and latter dismantled. The Death Valley region was primarily beneath a shallow, transgressing and regressing sea from some time prior to 1.7 BY to sometime in the early Jurassic (~200 MY), a period of up to 1.5 BY.

Beginning in the Late Paleozoic Era, by approximately 355 MY, oceanic tectonic plates converged with the western boundary of the continent of western North America and formed an active continental margin. Earthquakes, volcanoes, rising magma domes, mountain building, and compression forces folding or bending pre-existing rock resulted. This would continue intermittently until approximately 40 MY.

In the Mesozoic Era (251-66 MY), the environment would change from marine to a hot, terrestrial environment with periods of high humidity. Four periods of onrushing oceanic plate activity occurred during the Mesozoic Era creating Death Valley uplands, associated earthquakes, plutons, and volcanic activity. Finally, at approximately 40 MY in the Cenozoic Era, a relatively tranquil period of 25 MY would follow during which mountains would erode and a flat, savannah like environment result. A cooling trend began 33 MY. Tectonic forces began uplifting the Sierra Nevada early in the Cenozoic. The Sierra Nevada drained into the Death Valley region by the late Eocene (34 MY) and may have caused a rain shadow effect from the late Late Oligocene or early Early Miocene (24 MY) Epochs. A phenomenon termed "Great Basin Extension" would begin approximately 15 MY. This pulling apart of the Great Basin would produce its many basins and ranges. The Death Valley region saw mountains (Panamint, Grapevine, Funeral, Black, etc.) uplift and tilt while basins dropped. Warm and cold periods followed the Miocene Epoch with large mammals (e.g. mammoths, mastodons, horses, rhinoceros, etc.) roaming the Death Valley environs until a mass extinction approximately 10 KY. The hot dry desert environment did not begin until 7 KY.

So, what topography and rocks resulted from all this turmoil? **Figure 5** illustrates resulting mountains. They will be referred to in the following text.

A good overview of rock types is provided in **Figure 6**. We can see that the very old rocks (e.g. Precambrian gneiss) are generally in the lower, eastern Death Valley area along the west side of the southern Black Mountains, southern Nopah Range Alexander Hills, etc. The Paleozoic Era marine rocks (e.g. limestone and dolomite) run along the central spine and western edge of the Death Valley region. Late Precambrian rocks predominate in the west side of the Panamint Range. The Mesozoic Era intrusive (plutonic) rocks (e.g. granite,) can be found along the western edge of the Cottonwood Mountains, Argus and Slate Ranges, and southern Death Valley region (Owlshead, Granite, and Avawatz Mountains). Large areas of late Tertiary volcanic rocks (e.g. rhyolite, andesite, basalt) are scattered throughout the region. Tertiary Period plutonic rocks can be found in the southern Black Mountains and Kingston Range. Quaternary Period sand dunes and salts can be found predominantly along the central spine of Death Valley. A further appreciation of the complexity of the Death Valley region can be obtained by studying the stratigraphic column and maps in the Appendix.

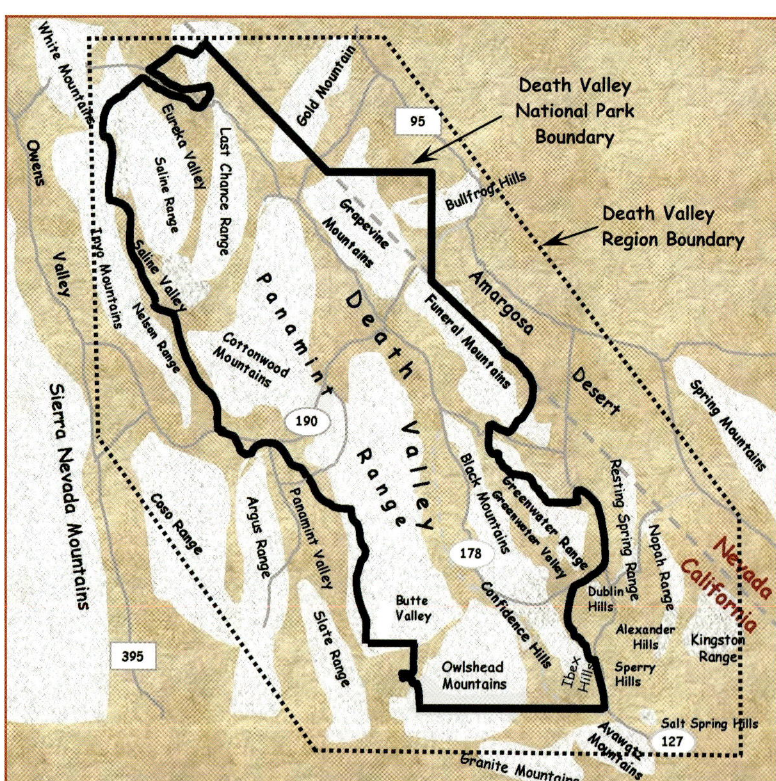

Figure 5 The Death Valley region has numerous mountains resulting from tectonic forces which uplifted and tilted ranges and down dropped basins.

Distribution of Rock Types

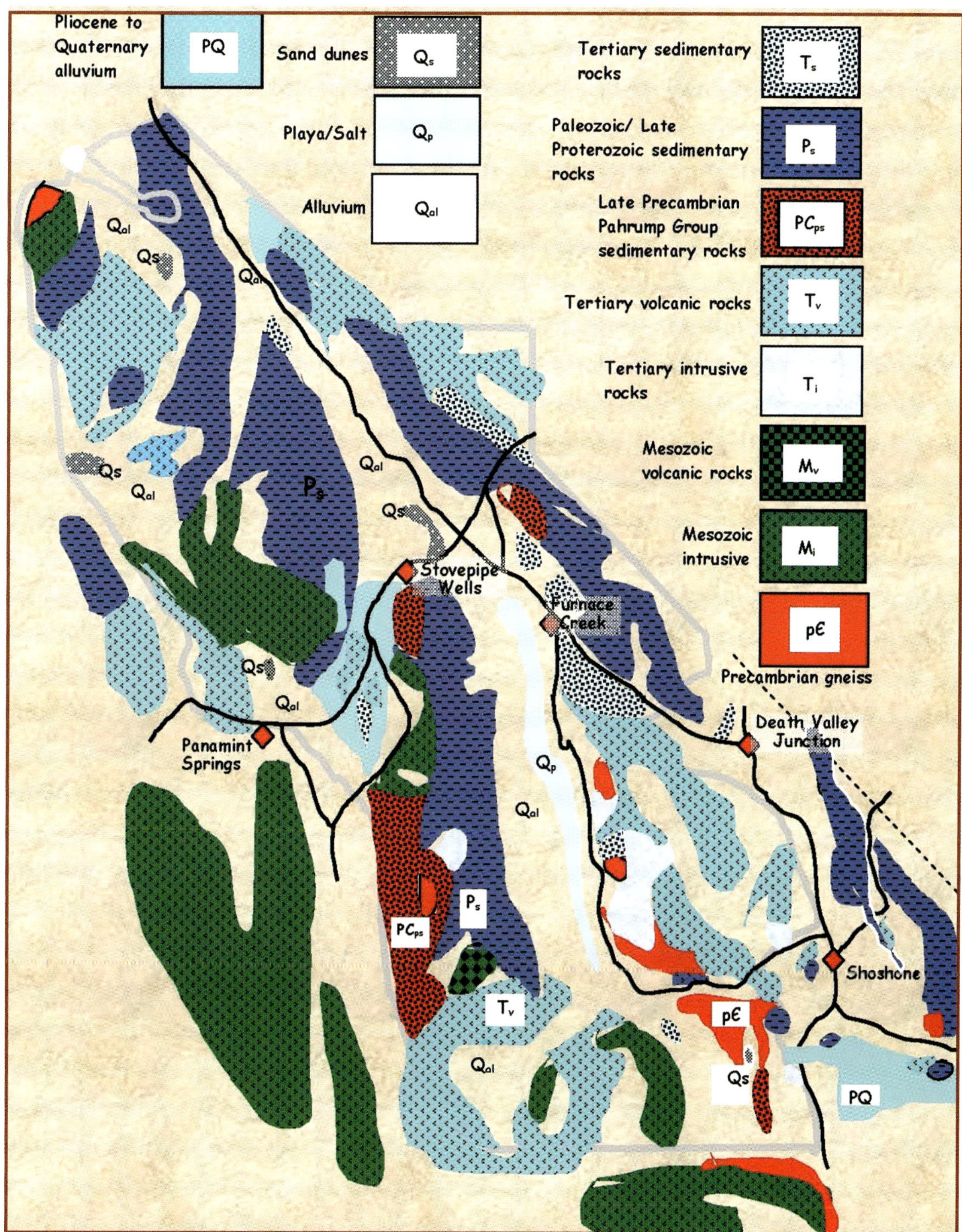

Figure 6 This sketch of the Death Valley region geology illustrates the wide diversity of rocks exposed as a result of uplift and tilting of the ranges.

Lengner, Reference: Marli Bryant Miller; USGS Topographic Sheets: Trona, Mariposa, Death Valley, Kingman

A Trip Through Death Valley's Geologic Past

The Death Valley geological history trip begins approximately two and an half billion years ago when the Death Valley region was on an ancient super continent that would someday be torn apart. The trip continues on to the present.

There are approximately 60 geologic formations (see Appendix) in the Death Valley region with the many formations containing numerous members. There are numerous ash falls originating in distant locales (e.g. Yellowstone, Bishop) and ash flows/falls and volcanic and plutonic activity from within the region. A sample from each of the aforementioned would yield myriad rocks which would tend to be repetitious and eventually confusing and meaningless to many. Selected rocks, which overview and highlight the region's history, will be discussed. Again, timeframe and sequence of deposition can be determined by text order and shading or via the Appendix.

Recall that there is a "physical" rock trail in Shoshone, California and that, Bennie Troxel selected samples representative of timeframes and lithology. This text provides a trip that can be considered to be a "virtual" rock trail. It uses many of Bennie's samples and many others as guideposts or slices in time. At each guidepost there are photographs and a discussion of the rock's composition, where one can see samples in the field today, and the ancient environment and life forms at the time the rocks were initially deposited.

The past and present come together as this brush lizard rests on a sedimentary rock deposited in a marine environment over a billion years ago.

T **This symbol is used throughout text to identify rock samples found along Troxel trail, Shoshone Museum.**

Hadean & Archean 4600-2500 MY
Setting the Stage

Proterozoic Eon 2500-542 MY
Ancient Continents and Beginning of Life

1.0 Crystalline Basement Rocks
2.0 Major Unconformity
3.0 Crystal Spring Fm.
4.0 Beck Spring Dolomite
5.0 Kingston Peak Fm.
6.0 Major Unconformity
7.0 Noonday Dolomite
8.0 Johnnie Fm.
9.0 Stirling Quartzite
10.0 Wood Canyon Fm.

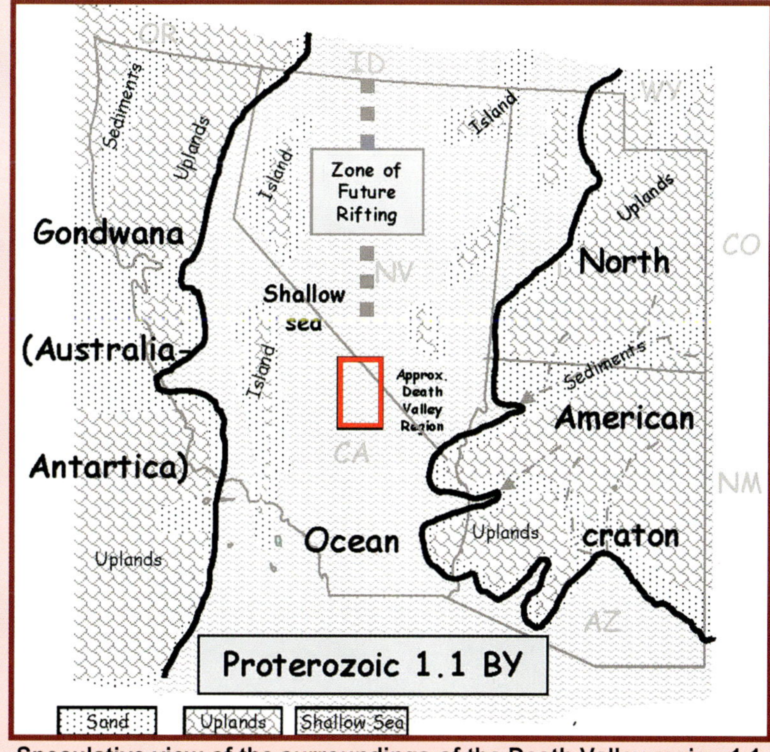

Speculative view of the surroundings of the Death Valley region 1.1 billion years ago. It is no longer on a continent (Columbia?) but beneath a shallow sea. *Reference Fiero, Scotese, Rogers, Santosh*

Hadean Eon (4,600-3,800 MY) and Archean Eon (3,800-2,500 MY) - Setting the Stage

During the Hadean Eon, Earth was heated by radioactive decay, asteroid impact, and gravitational condensation producing a molten core of heavy metals (iron and nickel) and a silica rich mantle. Earth's crust cooled, an atmosphere began, and the oceans filled. During the Archean Eon, continental and oceanic crusts were forming, the atmosphere continued evolving, and life on Earth began. Photosynthesizing cyanobacteria appeared (~3.5 BY) and began producing some of the oxygen that we breath today. No rocks of this timeframe are found in the Death Valley region.

Proterozoic Eon (2,500-542 MY)
Ancient Continents, Beneath the Sea, and the Beginning of Life

Death Valley was on the continent of Columbia at the beginning of the Proterozoic, but under a shallow transgressing and regressing (e.g. rising and falling) sea and in the southern latitudes by the end of the Proterozoic. Earth's continents were bare of life but the seas had various types of prokaryotes such as cyanobacteria which was prevalent, photosynthesizing, and creating oxygen for us to breathe. Later, in the early Proterozoic, single celled eukaryotes arrived (~1600 MY) but did not thrive and become the multicellular beginnings of animal life (ediacara) until the very late Proterozoic (~635-590 MY).

1.0 Earliest - Crystalline Basement Rock

[Early (Lower) Proterozoic: (2500-1700 MY)]

The "basement" rocks consist of gneiss and schist. Samples herein are gneiss. Gneiss is a rock that was formed deep underground when it was metamorphosed by intense pressure and temperature created by the overburden. The parent rock was volcanic, plutonic, and sedimentary. Gneiss is generally comprised of minerals such as feldspar (orthoclase), quartz, biotite, hornblende, and others. Chemical analysis of the gneiss of Death Valley's crystalline basement rocks indicates a date of 1.7 BY. Deposition was on an ancient continent named Columbia. Millions of years passed as the parent rock was covered by overburden and subsequently metamorphosed. Extrapolating back from 1.7 BY, it is estimated that the parent rocks were originally deposited on Columbia 2.5 BY. This gneiss can be found in the southern Nopah, Black, and Panamint Mountains.

Augen (German meaning eyes) gneiss sample. Large feldspar

Largest sample of Crystalline Basement gneiss shows some of the foliation of sample to right.

Foliated Gneiss. Planes resulting from flattening of constituent grains during metamorphism.

2.0 Major UNCONFORMITY (missing rock strata) in the Death Valley rocks between the Crystalline Basement and later Proterozoic marine rocks (~1.7-1.3 BY).

The Pahrump Group

(Crystal Spring Fm., Beck Spring Dolomite, Kingston Peak Fm.)

3.0 Crystal Spring Formation

[Mid Proterozoic (1300 MY ...)]

The rocks of the Crystal Spring Fm. (first of three in the Pahrump Group), were deposited during a time when the Death Valley region was transitioning from a terrestrial to a marine environment. An ancient continent was rifting and the Death Valley region was near the edge of a trough between the new continents. The rocks indicate a transgressing sea, followed by rapid uplift and erosion of the margin, and finally an immersion in a shallow transgressing and regressing sea. Rocks deposited during this timeframe include conglomerate, sandstone, dolomite, chert, diabase, and talc. The Crystal Spring Formation can be found in the Panamint, Kingston, and Nopah Ranges; Black and Avawatz Mountains; and Alexander Hills. The Crystal Spring Fm. has been measured at approximately 3,000 feet thick.

Conglomerates, Earliest (basal) of Crystal Spring Fm.

These samples (above and above right) are conglomerates from the lowest (earliest) member of the Crystal Spring Formation. The large, 30-pound sample of arkosic sandstone (right) has significantly rounded clasts indicating much prior clast erosion. Deposition may have been in a fluvial or a nearshore cobble beach environment.

Sandstone, Early (middle-lower) Crystal Spring.

The above, cross-bedded, feldspathic sandstone was deposited in a marine tidal environment. Wave action was responsible for the cross-bedding. Following deposition of said sandstone, the purple mudstone (right) was deposited. Lack of bedding planes may indicate a change to a low-gradient stream, estuary, or tidal marsh environment.

Mudstone, Early (upper-lower) Crystal Spring.

Stromatolites in the Crystal Spring Formation - Signs of Early Life, Enablers of Life

Signs of life are evident in the Crystal Spring Fm. in the form of stromatolites. The middle of the Proterozoic (~1,300 MY) was a time when there were abundant prokaryotes and eukaryotes. Indeed, prokaryotes dominated Earth's life forms from 3.5 BY to 600 MY. With no predation, prokaryotes prospered in the oceans. Cyanobacteria (a type of prokaryote), which created stromatolite fossils, populated tidal flats that bordered subtropical to tropical seaways where carbonate sediments were deposited. Other prokaryotes (methanogens, sulfur and heat loving bacteria and archea) also populated the oceans. Stromatolites are not the fossil remains of an prokaryotes, rather they are the layered structures formed by the trapping of sedimentary particles and precipitation of calcium carbonate in response to the metabolic activities and growth of mat-like colonies of cyanobacteria (prokaryotes) and some other prokaryotes. Some stromatolites look somewhat like mushrooms when viewed from the side. When cut perpendicular to the longitudinal axis (top view), they look like deformed concentric circles. The accompanying photos show a side and top view of some Crystal Spring Fm. stromatolites.

Top view of stromatolites appear to be deformed concentric circles.

Side view of stromatolites in the Crystal Spring Fm., Alexander Hills.

Stromatolites can also be found in ensuing Death Valley rock formations such as the Beck Spring Dolomite, Johnnie Fm. and Noonday Dolomite. They lost their grip on world domination ~635-590 MY with the advent of the first animals...ediacara which left traces in the Wood Canyon Fm. In Australia's Hamlin Bay, stromatolites are still being formed today.

It is important to note that cyanobacteria and other photosynthetic bacteria produced oxygen. They changed Earth's atmosphere from one that would have been toxic to subsequent life forms such as dinosaurs and ourselves. These prokaryotes can be considered to be the enablers of life as we know it.

Stromatolites forming in Hamlin Bay, Australia.

Looking like giant mushrooms, these stromatolites found in Hamlin Bay, Australia are the current version of the stromatolites that first appeared ~3.5 BY. Current stromatolites are found in warm, humid climates in shallow, subtidal-intertidal environments possibly extending to supratidal environments. They can occur in freshwater. *A. J. Frank*

Dolomite, Upper Crystal Spring.

Algal Mats, Upper Crystal Spring.

Above are two samples of the Crystal Spring dolomite deposited in the shallow sea covering the Death Valley region. Dolomite is limestone (calcium carbonate) that has some of the calcium element replaced by magnesium. It can be referred to as calcium-magnesium carbonate. The large tan sample to the left has a smooth, sandstone grittiness with what appears to be holes. When the limestone was originally deposited, it was not homogeneous. Softer sediments eroded out of the dolomite leaving behind the "holes." On the left side of that sample is coarse black/brown mounds of chert. Chert is another sedimentary rock that is silica oxide (like flint or jasper). The smaller sample to the right shows distinct beds or layers of harder and erosion resistant chert protruding above the layers of softer limestone or dolomite. This dolomite/limestone and chert is what was metamorphosed into talc by the upwelling hot diabase. The chert would have been deposited in a shallow, tidal flats environment.

1,100 MY Divergent Boundary (Rift), Diabase Intrudes Pre-existing Crystal Spring Fm.

Diabase is a plutonic rock that results when magma, consisting of feldspar (plagioclase) and pyroxene minerals, intrudes pre-existing rock. Primarily black, it often appears to have a dark greenish cast. Chemically it is approximately half silica oxide (SiO_2), 15% aluminum oxide (Al_2O_3) and 15% iron (Fe). Approximately 1.1 billion years ago a rift formed and magma rose and intruded the pre-existing dolomite of the Crystal Spring Fm. The heat from the magma "cooked" the cherty dolomite via contact metamorphism and produced tremolite and commercial talc. The diabase occurs in sills from tens of feet to greater than 1,500 feet thick. Diabase locations include but are not limited to the Kingston and southern Panamint Ranges, Alexander and Ibex Hills, and Owlshead Mountains.

Black diabase that intruded Crystal Spring Fm. and formed talc.

Commercial Talc, from Intrusion of Diabase 1,100 MY

Whitish, Soft Talc, East Side of Southern Panamint Range.

1,100 MY hot diabase changed the cherty dolomite parent rock into commercial talc by contact metamorphism. Commercial talc constituents include tremolite, formed nearest the hot diabase, and talc. Both are comprised of magnesium, silica, oxygen, and hydrogen; but, tremolite has calcium and iron added. Laminated and massive tremolite is white (sometimes green), fine grained, feels soapy, and is relatively soft but harder than talc. Tremolite cleavage is in two directions like a diamond (124^0 and 56^0). Talc (magnesium silicate), or talc schist, is apple green, white or pearly, has perfect cleavage in one direction, is fine grained and soft, and feels greasy. Commercial talc is used in ceramics, plastics, and lubricants. It is not the talc that was once used on babies. Talc bearing areas include but are not limited to the southeastern Panamint, southern Nopah, and Kingston Ranges; southern Black and northern Avawatz Mountains; Ibex, Saratoga, Saddle Peak, Alexander, and Silurian Hills; and Silver Lake area. The above sample has been weathered and lost some characteristics of freshly exposed talc.

Western Talc mine in the early 1970s.

Western Talc mine in the Alexander Hills was one of the first talc mines to open (~1911) in the Death Valley region and one of the last to close (late the 1970s). Photo to left shows the Western Talc shortly before closure. To the top left of the photo is the main shaft. Underground, the miners extracted the talc and used hoists to get it to the conveyor belt in the center of the photo. The talc was dumped into an ore bin where it was graded and from which another conveyor lead. Outside of the photograph, to the right, was a series of concrete ore bins. Depending on the quality of the ore, the second conveyor was moved to one of the concrete ore bins where the ore was dumped and stored for pickup. Trucks then transported the ore down Sperry Wash to Dunn siding on the Union Pacific Railroad.

Underground in the Big Talc mine, southern Panamint Range, circa 1946-1951. (l-r) Louise Grantham, miner, Ernie Huhn, and miner. Miner in white shirt is "slushing." Talc is scooped up and drug up ramp until in falls into an ore car in front of the slusher. Note the pulley on the ceiling. Another pulley was near the talc loosened by dynamite or other means.

Wollastonite from Intrusion of Diabase 1,100 MY

Wollastonite, Southern Panamint Range.

Wollastonite is a product of contact metamorphism of impure limestones. This 2' long sample is from the southern Panamint Range in Warm Springs Canyon, near extensive talc mines once owned by Ernie "Siberian Red" Huhn and Louise Grantham. Wollastonite is white-grayish, has the luster and appearance of glass, feels silky, and its cleavage is at right angles while tremolite cleaves at 56^0 and 124^0. It is harder than talc but is nearly of the same hardness as tremolite. It is fluorescent.

Iron a Constituent of Diabase 1,100 MY

Iron Ore from the Kingston Range.

Diabase is approximately 15% iron. High in the Kingston Range, a concentration of iron (50 pound sample to far left) was mined for many years. In this iron ore, many deposits of white calcite and "Fool's Gold" (iron sulfide), can be found (see below). Some of the ore is magnetized and correctly referred to as "magnetite."

Fool's Gold and Calcite in Iron Ore.

4.0 Beck Spring Dolomite

[Mid Proterozoic Eon (<1,100- MY ...)]

Beck Spring Dolomite is the second in the Pahrump Group. It is uniformly grayish and massive (no flow layering, foliation, cleavage, joints, fissility, or thin bedding planes, etc.) to well-bedded, cliff-forming, and approximately 1000 feet thick. Diabase, prevalent in part of the Crystal Spring Fm., is not evident in the Beck Spring Dolomite. Numerous stromatolites and small, round pisoliths (formed by algal encrustation and rolling on sea floor) in the Beck Spring help to identify a shallow marine depositional environment. The Beck Spring is found in, but not limited to, the southern Black Mountains; Kingston Range; and Alexander, Saratoga, Saddle Peak Hills. In the Panamint Mountains it is identified as the Marvel Limestone.

Beck Spring Stromatolite. View is looking down on top of the stromatolite. Alexander Hills.

This large Beck Spring sample is massive and medium gray. It does not contain stromatolites or pisoliths seen in the other pictures which were taken at various locations in the Alexander Hills and Saratoga Hills.

Columns of layered stromatolites climb this out crop of grey Beck Spring Dolomite in the Alexander Hills.

The pisoliths in this photograph are the same size as they were close to a billion years ago when they rolled around near the shore of an ancient sea. Saratoga Hills.

 **This symbol is used throughout text to identify rock samples found along Troxel trail, Shoshone Museum.**

5.0 Kingston Peak Formation

[Mid Proterozoic Eon (<1,100- MY ...)]

The rocks of the Kingston Peak Fm. (KP) , third and upper formation of the Pahrump Group, can be found in Death Valley region's Funeral Mountains; Panamint and southern Nopah Ranges; Alexander and part of the Saddle Peak Hills; and what we will term the southern Death Valley region. Troxel has divided the southern Death Valley region into an area (southern Black Mountains, Kingston Range, Sperry Hills) where the northern facies of the KP can be found and an area (Saddle Peak, Silurian, and Salt Spring Hills) where the southern facies of the KP can be found. Generally, the KP tends to be reddish with the exception of the southern facies which is dark gray to black. In the Kingston Range, the KP is up to 10,000 feet thick. The depositional environment of the KP varied from a marine nearshore to deeper marine environments.

KP Northern Facies, Southern D.V.

Deposits from northern source have a reddish matrix, angular clasts of gray dolomite and feldspar with lesser amounts of quartzite, diabase, gneiss as seen below.

The above KP sample is diamictite (coarse, angular to well-rounded clastic fragments supported in a fine matrix), from the middle of the northern facies of the KP. Diamictites can be deposited by various means including glacial till, ice rafted sediments, debris flows, turbidity currents, etc. Most feel the origin of the KP is glacial. Some feel that this clay, silt, sand, gravel, and very different sizes of boulders was deposited directly by a glacier. Their evidence is rare stones with striations (multiple, parallel scratches) and even rarer stones with facets (plane surface caused by abrasion). Troxel's years of analyses indicates a different origin. He believes that, "these features are derived from a distant glacial terrain but the layers in which they are contained were deposited in submarine fans - so called drop stones may well be carried into the area of deposition by icebergs but may have been dropped many miles from the point of origin of the icebergs."

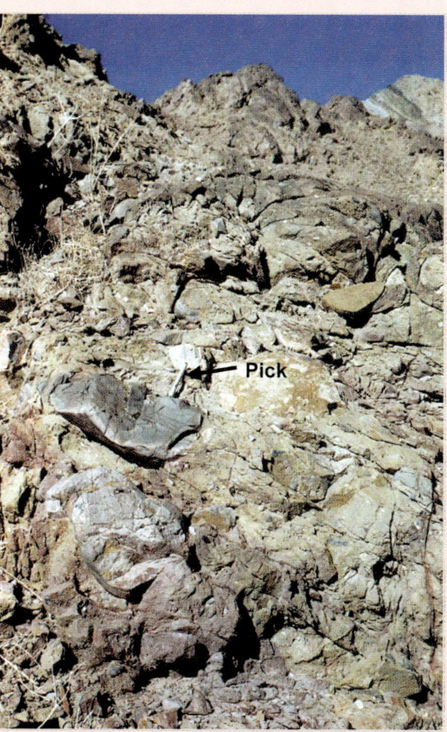

Outcropping of northern facies diamictite with huge clasts, southern Black Mountains. See geologist pick, center photo for scale.

The above 2-foot long, KP sample has gravel-sized, slightly rounded breccia from the Crystal Spring (e.g. feldspars), Beck Spring (gray dolomite), and other pre-existing sources that continually grade finer to the red arkosic sandstone seen on the left. It is from the upper KP and from a nearer shore environment than the diamictite.

This close up of the slightly rounded breccia seen in photo to the left shows the pre-existing rocks that comprise the graded beds.

KP Southern Facies, Southern D.V.

Southern facies is dark gray to black and consists primarily of subangular gneissic clasts derived from a southern source area.

Bennie Troxel near the Saddle Peak Hills on his way to study the southern KP.

The above samples are from the southern facies diamictite. Deposited from a source to the south, the matrix is gray and the clasts are primarily granite, quartzite, and gneiss. Deposition method is basically the same as the northern facies. In the southernmost Saddle Peak Hills, the northern and southern facies interfinger.

Large outcropping of southern facies of Kingston Peak Fm. diamictite in the Saddle Peak Hills, Death Valley National Park.

This symbol is used throughout text to identify rock samples found along Troxel trail, Shoshone Museum.

6.0 Major UNCONFORMITY (missing rock strata) in the Death Valley rocks between the Kingston Peak Fm. and later Proterozoic Noonday Dolomite

7.0 Noonday Dolomite
[Late Proterozoic (~1,000-700 MY)]

The Noonday mine intermittently produced lead and silver from 1878 to 1928. The standard gauge Tecopa Railroad hauled ore from 1910 to 1922 from the mines (Noonday and Gunsight) to a mill and juncture with the Tonopah & Tidewater RR.

The Noonday Dolomite was named after the Noonday lead-silver mine (left). It was deposited in a marine environment before animal life swam in the seas or plants and animals populated the land. Eukaryotes would not evolve into multi-cellular animal life (metazoans) until ~635 MY. Exposures of the Noonday Dolomite can be found in the Panamint, Resting Spring and Nopah Ranges; Avawatz and southern Black Mountains; and Alexander, Salt Spring , north Sperry, Saddle Peak, Ibex, Saratoga, and Dublin Hills. In the northern exposures of the Noonday, it is comprised of two dolomite members while in the southeast a third member, known as "basin faces," can be found. The Noonday's two dolomite members are greater than 1,200 feet thick. To the south-southeast of the Noonday distribution, there is an abrupt disappearance of the dolomite members and appearance of the clastic (fragments of rocks) units of the basin facies.

The sample above is from the lower dolomite layers of the Noonday Dolomite. Deposition was in a shallow marine environment. Some of the dolomite is layered and some is not. Outcroppings are pale grays and tans. A prominent feature of this sample is what appears to be holes. In actuality, they are the ends of tubes once rising vertically through the layers of algal mats. An unproven theory is that these tubes vented gases from the decaying algal masses below. If one were to hypothetically step away from the sample, it may appear to be within a series of concentric circles termed stromatolites. Sticky secretions from the algae trapped sediments and formed the circular (in plan form) stromatolites. Large, concentrically banded stromatolitic mounds are found in the southern Nopah Range.

The above diamictite is from the lower basin facies of the Noonday and contains angular fragments of limestone and dolomite. The sample has weathered a yellow-brown. It was deposited in a deeper marine setting than the Noonday's dolomite.

Road leading to a huge algal mound in the Noonday Dolomite, southern Nopah Range.

8.0 Johnnie Formation

[Late Proterozoic (~900-800 (?) MY)]

The Johnnie Fm. can be found in the Spring Mountains and Nopah, southern Panamint, Resting Spring and Kingston Ranges. It is conformably above the Noonday Dolomite and below the Stirling Quartzite. The Johnnie is up to 4000 feet thick. It forms dark colored slopes. Its members include various manifestations of dolomite, quartzite, shale, and siltstone with iron present in some dolomite. Stromatolites and oolites (concentric carbonate laminae surrounding a nucleus [e.g. quartz]) are present. Although there are exceptions, generally, it grades finer upwards indicating a transgressing sea that is steadily getting deeper. There were still no plants or animals on land. In the seas, eukaryotes had not yet evolved into multicellular life. Much later they would later evolve into plants, animals, fungi, and protists.

This is basal (the lowest portion) Johnnie Fm. It is, cross-bedded, siliceous (containing abundant silica) dolomite indicating a high energy shoreline environment.

Crossbedded sandstone from the lower Johnnie Fm.

The lower rock is a purple siltstone formed in a calm, near shore environment. It has non-directional ripple marks. Irregular bedding is seen on its sides. Resting upon the siltstone is tan, sandy dolomite containing numerous oolites deposited in agitated water in depths measured in feet (~5') and not yards.

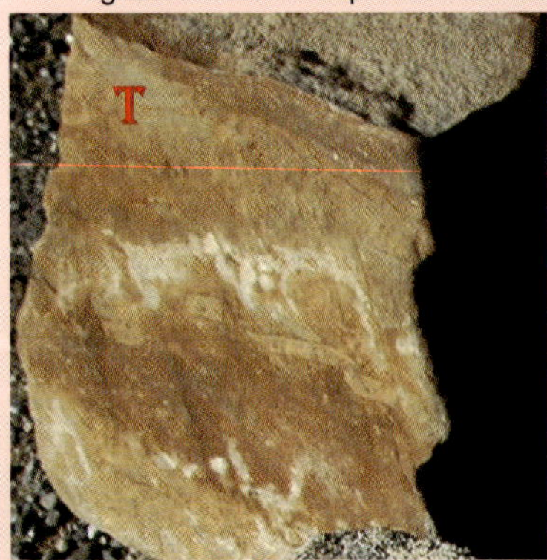

Siltstone of upper Johnnie Fm. deposited in deep water.

Partially metamorphosed Johnnie Fm.

Johnnie Fm. quartzite pebbles in sandstone matrix later metamorphosed into quartzite.

9.0 Stirling Quartzite

[Late Proterozoic (~800-700 (?) MY)]

The Stirling Quartzite is widely exposed throughout the Death Valley region including the Avawatz, southern Black, Grapevine and Funeral Mountains; Kingston, Nopah, and Panamint Ranges; and Salt Spring and Alexander Hills. The Stirling is up to 2,000 feet thick and forms yellowish or pinkish slopes. The formation was deposited in marine (passive continental margin) and terrestrial environments.

The arrangement of its members are such that the upper and lower parts of the Stirling are almost mirror images of each other. Characteristics such as, "bed thickness, mean grain size, and resistance to weathering tend to decrease, and the proportion of siltstones and amount of potassium feldspar tend to increase up-section from the bottom to the middle." From the middle to the top of the Stirling, these trends are reversed. The upper Stirling terrestrial deposit is dark and fine-grained while displaying mud cracks, rain drop impressions, and ripple marks from paleo currents.

This sample is not pure quartzite. It is from cross-bedded sandstone (with feldspar) formed near shore, tidal environment. It appears to be from the upper Stirling.

Marine deposits display coarse-grain-size-to-fine followed and then back to coarse-grains. This indicates a high energy to low energy and back to high energy environment. Deposits were in a tidal system. Water depth got deeper up to the middle of the Stirling depositional sequence and then got shallower. Multicellular sea life was progressing; however, the first animals would not populate the seas until approximately 635-590 MY. The land was still barren of any life. The continents had all drifted to the far southern latitudes.

This photograph of part of the middle member of the Stirling was taken near Grimshaw Lake near the town of Tecopa, CA. Scale is provided by the hat at the bottom of the photograph. This micaceous siltstone is grayish-purple and shines in the sunlight. Note the asymmetrical ripple marks. Along with the weather resistant, fine grains of the siltstone, they denote a near shore depositional environment.

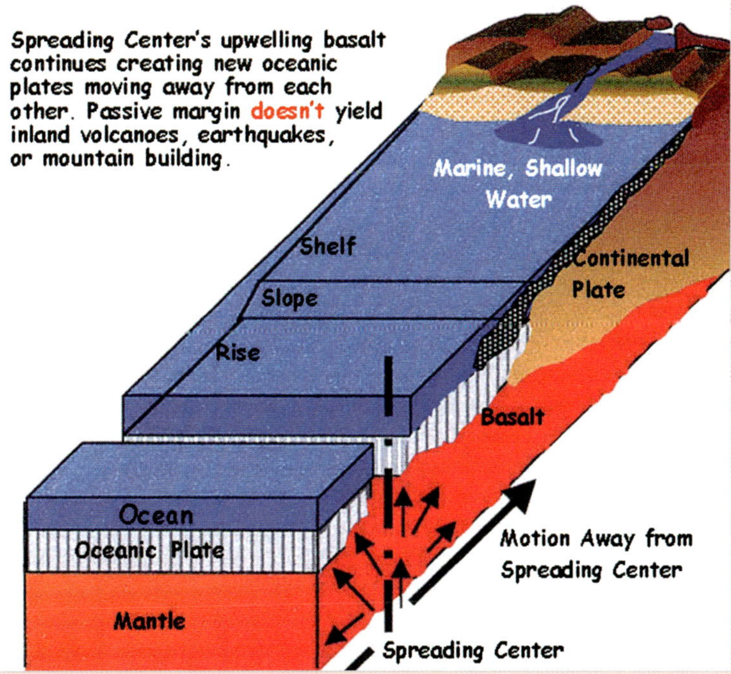

By the Late Proterozoic Eon, the Death Valley region had been carried far from the original 1.1 BY diverging boundary and was on a passive continental margin (illustrated above) where it spent the majority of the Paleozoic Era. Sediments along the shallow passive margin included limestone and dolomite derived from a marine environment and terrestrial sediments of sandstone, conglomerate, and siltstone that were derived from inland sources such as rivers or deltas.

Transition: Proterozoic Eon to Paleozoic Era - Animal Life Blossoms

10.0 Wood Canyon Formation

[Late Proterozoic-Early Cambrian (~700-500 MY)]

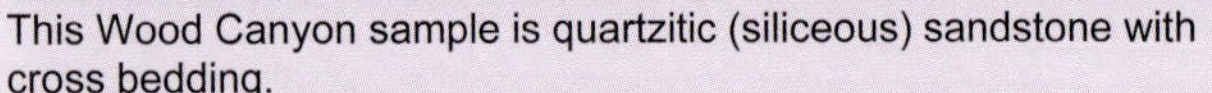

This Wood Canyon sample is quartzitic (siliceous) sandstone with cross bedding.

Top photo is spiral mound and lower photo is of worm burrows from upper Wood Canyon, shaly siltstone.

The Wood Canyon Fm. is found in the Kingston, southern Nopah, Panamint, Last Chance, and Resting Spring Ranges; Grapevine and Owlshead Mountains; and the Alexander and Salt Spring Hills. It lies conformably between the Stirling Quartzite below and the Zabriskie Quartzite above. The Wood Canyon is approximately 1,700 feet thick.

One of the interesting things about this formation is its diverse lithologies which include a multiplicity of sandstone, siltstone, conglomerate, shales, quartzite, and dolomite with colors and shades including variations on gray, red, brown, tan, yellow, maroon, pink, orange, green, purple, and olive. This diversity has been grouped into three members. The lowest member (~265') is comprised of siltstone, sandstone, and dolomite with symmetric ripple marks, mudcracks, cross bedding, and possible worm burrows in the upper portion. The middle member (~835') begins as an arkosic conglomerate with the final three quarters grading finer into numerous cycles of colorful quartzitic, feldspathic, and arkosic sandstones and maroon, red, purple, etc. siltstones. Cross bedding occurs throughout the member with worm burrows in the upper portion of the middle member. The upper member (~600') is comprised of numerous beds of sandstone and siltstone with ubiquitous cross bedding. What sets the upper member apart from the rest, is the numerous early Cambrian fossils found therein. Fossilized worm tracks are followed by small, free swimming, marine gastropods; trilobites; brachiopods; echinoderms; and spiral molds.

The second interesting thing about the Wood Canyon is the large number of the aforementioned fossils which happen to document some of the earliest animals (ediacara) and the beginning of the first predatory and skeletal animals. The Wood Canyon sits athwart the boundary between the Proterozoic Eon and the Paleozoic Era's Cambrian Period (542-488 MY). (Return to the geologic time scale, page 3.) At that time, the region that was to become Death Valley was offshore from a continent that is now called Laurentia which was part of supercontinent Pannotia. This continent and its coastal waters were on a tectonic plate that had been drifting north from its presumed location in the southern latitudes. It was not at the equator but it would reach and pass it before the end of the Cambrian. The earth was cooler than it is today. Inland, there was no plant or animal life, only the barren landscape and some glacial ice sheets. It would be nearly another 100-million years before the first land plants would take root, 200-million years before the first amphibians crawled ashore, 400-million years before the age of dinosaurs would be in full swing, 515-million years before the age of mammals would begin, and approximately 580-million years before *homo sapiens* would build the Death Valley Visitor Center at Furnace Creek. The continent's passive margin was quiet with little or no volcanic activity or earthquakes. The Southwest had been both above and below the ocean waters due to early cooling and subsidence of the continental margin, sea floor spreading rates, or rising and falling of world wide sea levels. The Wood Canyon Fm. is primarily shallow tidal deposits with occasional terrestrial deposits.

During Wood Canyon Fm. deposition, Earth's seas did not teem with fish as they do today. Instead, life had recently evolved from single cell eukaryotes to multicelled life (metazoans)... the first animals called ediacara. The ediacara had no bones and lacked teeth or any other offensive weapons with which to attack their fellow metazoans. They left some late Proterozoic fossils in Death Valley National Park's Boundary Canyon and in the southern Nopah range. Some looked like jellyfish, huge blobs, fried eggs, threads, pancakes, ribbons, worms, or plantlike organisms. The sea was full of weird shapes. Some floated while others may have been attached to the bottom. After ~20 MY, the ediacara, in turn, disappeared abruptly near the early Cambrian which saw animals that burrowed, had skeletons, and were predatory. Trilobites, echinoderms (radial symmetry, e.g. starfish, sand dollars, sea urchins) and brachiopods (bottom dwelling, opposing shells not symmetrical like clams) were early Cambrian arrivals.

Paleozoic Era (542-251 MY)

Beneath the Sea, Ancient Life

11.0 Zabriskie Quartzite
12.0 Carrara Fm.
13.0 Bonanza King Fm.
14.0 Eureka Quartzite
15.0 Tin Mountain Limestone

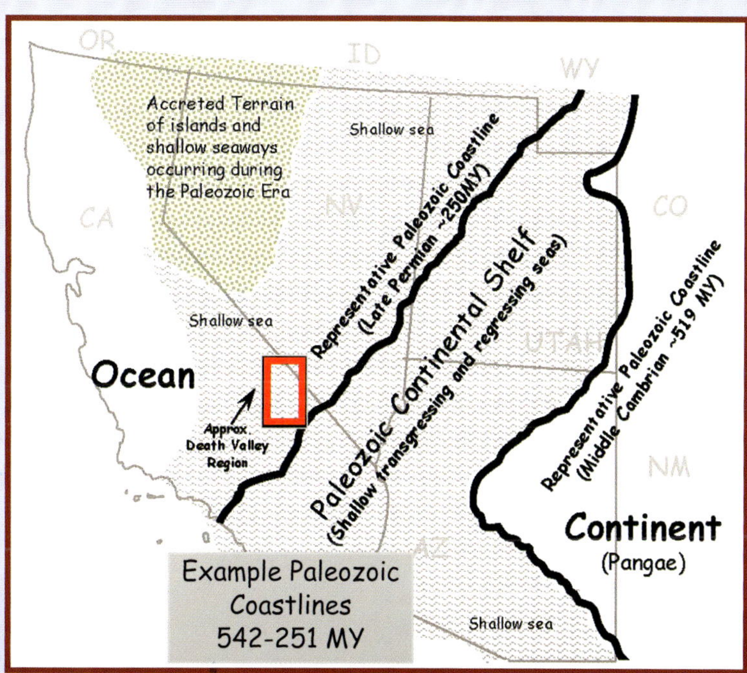

During the Paleozoic Era, the Death Valley region migrated from the far southern latitudes to northern latitudes. It was on a passive continental margin, AND beneath a shallow transgressing-regressing sea (exception of a brief period during the late Cambrian and the Ordovician Periods). Numerous thick layers of carbonate rocks were deposited throughout southwestern North America. Shown here are a few representative coastlines. The continental margin would become active in the Late Paleozoic. Death Valley would not emerge permanently from the waves until the Late Triassic-Early Jurassic Period.

With the exception of dips in the Late Ordovician and Pennsylvanian Periods, global temperatures were warmer than today. At the end of the Paleozoic Era, plant and animal species suffered Earth's worst mass extinction.

Paleozoic Era (542-251 MY) - Beneath the Sea, Ancient Life

The beginning of the Paleozoic Era saw seven major continents surrounded by the Panthalassic and Prototethys Oceans. The Death Valley region was beneath a shallow, transgressing and regressing sea, on a passive continental margin, and near the equator. By the end of the Paleozoic Era, there was only the supercontinent called Pangea which was surrounded by the Pacific, Panthalassic, and Tethys Oceans. During the intervening timeframe, the Death Valley region had witnessed and recorded innumerable transgressions and regressions of the shallow sea that covered it, the continental margin became active because oceanic tectonic plates were converging on the western North American coast, and Pangea had drifted into the far south latitudes.

Earth's Paleozoic Era atmospheric oxygen content eventually trended higher than today's and provided an impetus for plant and animal growth. Temperatures were initially higher but ended up lower than today's. The land, which was essentially bare of plants at the beginning of the Paleozoic Era, evolved into the home of forests and coal bearing swamps with club mosses, horsetails, ferns, cycads, seed ferns, primitive conifers, and ginkgoes. Arthropods (millipedes, scorpions, mites, spiders, and insects) led the charge to occupy the land. Eventually large, carnivorous, egg laying reptiles like the fin-backed *Dimetrodon* would prowl the land. Herbivores like *Lystrosaurus, Kannemeyeria*, and head butting *Moschops* dodged the predators. The seas started out with early animals that developed internal and external skeletons and experimented with body forms until "agreeing upon" today's body forms. Numerous fish (sharks, ray-finned, etc.) and invertebrates (foraminifera, brachiopods, crinoids, corals, bryozoans) evolved. A major mass extinction at the end of the Paleozoic was triggered with the basaltic eruptions of the Siberian Traps. Over ninety percent of life was wiped off the face of Earth.

11.0 Zabriskie Quartzite

[Early Cambrian (~500 MY)]

The Zabbriskie Quartzite can be found in the Grapevine and Funeral Mountains and the Panamint, Nopah, Resting Spring, Last Chance, and Kingston Ranges. Sequentially, it is beneath the Carrara and above the Wood Canyon Formation. Its thickness has been reported as much as 930' in the Grapevine Mountains and as little as 150' in the Resting Spring Range. It weathers a dark reddish-brown or purplish but when broken up it glistens almost white or pink. Its sandstone is course to fine grained with well-rounded grains. Sandstone or quartzite bedding is laminated or very thin with some cross bedding. Some mudcracks and raindrop impressions have been reported. Current (symmetrical) ripple marks are found in the top of the formation. It splits flaggy, forms ledges, and is more erosion resistant then the Wood Canyon Formation beneath it. The Zabriskie was deposited on a broad, shallow, passive, continental margin in a low-energy nearshore or an intertidal (generally between high and low tide) environment.

This sample (upper left) of the lower Zabriskie consists of: fine grained sandstone with worm tubes overlain by yellow gray parallel laminated siltstone. The end view of worm tubes can be seen in the photo above. The photo directly to the lower left is of the upper Zabriskie Quartzite which consists of purple, pale red, and grayish orange laminated layers of siltstone or quartzite with occasional traces of trilobite fragments or trace fossils. Fractured samples of this purple Zabriskie sparkle in the sun and are one of the most visibly esthetically pleasing rocks in the region.

12.0 Carrara Formation

[Early Cambrian (~550 MY)]

The Carrara Fm. can be found in the Grapevine and Funeral Mountains, Panamint, Nopah, Resting Spring, Last Chance, Kingston, and southern Argus Ranges. Thickness has been reported as 1,530' in the Grapevine Mountains. It can be seen between the distinctly striped and banded Bonanza King Fm. and the dark red-purple-brown Zabriskie Quartzite. Generally, the upper, approximately 65% is grayish limestone/dolomite that forms steep, gray and light brown slopes. The lower 35% is primarily olive, green, or brown siltstone with evidence of stromatolites, oncolites, archaeocyathids, oolites and trilobites. The lower Carrara was likely deposited in tidal flats whereas the upper may have been deposited farther offshore.

This tan and gray, weathered limestone sample of the Carrara Fm. has brownish chert (once algal mats) and is coarsely weathered with deep, angular furrows with sharp edges. In the Death Valley environment, limestone and dolomite weather with these furrows and sharp edges which can severely cut an unwary or clumsy hiker's hands, legs, arms, etc.

These oncolites, found in the Carrara Fm. in the Nopah Range, once rolled around along the early Cambrian shoreline. They were round, layered masses of algae.

This excellent trilobite fossil was preserved for posterity in the siltstone of the Carrara Fm. It once crawled upon the early Cambrian sea floor as seen in the accompanying painting.

Trilobites are crawling along the sea floor looking for prey or something to scavenge. The crustacean *Canadaspis*, which extracted small food particles from sediment, is shown swimming (upper left). Sponges, present since the Precambrian, are to upper right. Although attach points and bodies are not clearly seen, early crinoids (marine animals) may be represented by the two types of slender, seaweed like depictions in the background. *Heinrich Harder*

13.0 Bonanza King Formation

[Middle - Cambrian (~520 MY)]

The Bonanza King Fm. (BK) is widely distributed throughout the Death Valley region including the Inyo, Funeral, and Grapevine Mountains and the Last Chance, Panamint, Argus, Nopah, Resting Spring, and Kingston Ranges. The BK is roughly 3,600-4,000 feet thick and consists of interbedded limestone and dolomite with occasional sediments from the continent. Sequentially, it is between the Nopah Fm. (above) and the Carrara Fm. (below). The formation forms steep, weather resistant slopes and appears to be composed of wide, horizontal stripes that are different shades of gray and occasionally some shades of brown, as seen in the photo below. Due to impurities, both limestone and dolomite can be different shades of gray. Dolomite is slightly more resistant and may appear to bulge outward or form ridges when interbedded with limestone.

The BK was deposited in the Middle to early Late Cambrian (~530 to 515 million years ago). The Death Valley region was still on a passive continental margin covered by a warm, shallow sea, saturated with calcium carbonate. As in the past hundreds of millions of years, the seas transgressed and regressed due to glacial events, margin subsidence, or sea floor spreading rates. The major source of the Bonanza King limestone stems from a time when the depositional environment was offshore, slightly farther than the intertidal zone and hence, relatively calm waters. A muddy calcite seafloor resulted in limestone that was finely textured and thinly bedded. When the seas retreated, an intertidal environment could result and yield stromatolites and thin limestone beds. If the depositional environment were closer to the shoreline, a coarser limestone would result. The mottling found in the lower portion of the BK is due to seafloor creatures that burrowed and tunneled their way through the mud. Some oolites and pisoliths are also present in the lower BK.

This sample is from the lower Bonanza Kin Fm. and displays dark and light gray mottling. It is due to seafloor creatures that burrowed and tunneled their way through the half billion year old mud. This process is called "bioturbation".

The Bonanza King Fm. can easily be spotted due to it's alternating layers of dark and light gray dolomite and limestone. Photo is from Titus Canyon, west of Leadfield.

The Cambrian Period (542-488 MY), witnessed by the BK and its predecessors (Wood Canyon, Zabriskie Quartzite and Carrara), was an extremely important timeframe for animal life on Earth. Animal life in the seas developed all animal body plans we have today and even discarded a few. Perhaps it is best summed up as follows:

> *"Over those roughly 50 million years, the vast majority of animal phyla first appeared. All specialists agree that this is thee most important event in the entire history of animal life, superseded in importance only by the first appearance of life on Earth, perhaps, in the context of the entire history of life on our planet."*
>
> *Peter D. Ward, Out of Thin Air, 2006*

As for the land, things were different. Approximately 520 MY, there was traces of green in wetter areas but most of the surface area was still bare rock. There were no trees, bushes, flowers, grasslands and no animals. There were no fish in the streams, no insects, no amphibians, no wading birds, and no dragon flies. There was a low atmospheric oxygen level... too low for us. Things changed.

14.0 Eureka Quartzite

[Middle Ordovician (~465 MY)]

The Eureka Quartzite can be found in the Last Chance, Panamint, Argus, Nopah, and Resting Spring Ranges and Grapevine and Montgomery Mountains. Total thickness is estimated at 350 feet. It uncomfortably overlays carbonate beds. The upper part is quartzite that is distinctly white, has the luster and appearance of glass, has some cross-bedding, and fractures a sparkling white. The source of this very white quartzite was very clean (no impurities) sands possibly provided by long shore currents. Lack of fossils indicates a subsequent regression (end of early Ordovician) with dry land exposed. The lower portion is hematitic (iron) platy quartzite with sandy and silty dolomite and limestone. The stromatolites in the lower portion bespeak a nearshore, tidal environment present before the regression. In the middle Ordovician Period, the Earth's sea levels would rise again.

The distinctly very white, upper portion of the Eureka Quartzite.

The Eureka Quartzite was deposited during the middle of the Ordovician Period (488-444 MY) when the land was greener than in the Cambrian and the equator ran through what is now northern California. There were mosses and primitive vascular plants (bryophytes) present; however, much bare rock was still visible on land. There were still no trees, flowers, shrubs, insects, birds, or any animals on land; however, the sea was a different story because there were almost three times as many marine animal families at the end of the Ordovician Period as at the end of the Cambrian Period. In the shallows of the sea, there were many different sponges. A new animal arrival, colonial bryozoans, looked somewhat like a collection of miniature brachiopods. There were colonial corrals (tabulates); large calcareous, extinct sponges (stromatoporoids); and solitary horn shaped corals (rugose). No schools of fish inhabited the reefs; instead, swimming and crawling trilobites and other arthropods occupied the reefs. Having surpassed the arthropods as top predators, the new cone-like cephalopods were present in abundance. Heavily armored, jawless fish (ostracoderms) with sucker like lips fed on surface algae. A mass extinction at the end of the Ordovician Period, possibly due to global cooling and anoxia, caused a loss of 60% of marine genera.

15.0 Tin Mountain Limestone [Early Pennsylvanian(~350 MY)]

Tin Mountain Limestone with chert deposits.

The Tin Mountain Limestone can be found in the Inyo Mountains and. Last Chance, northern Panamint, and Argus Ranges. Estimated thickness is 1000 feet. It is a bluish-gray, fine-grained, cherty limestone. The lower is thin-bedded while the upper is thick-bedded. It was deposited in shallow marine, offshore bar, lagoon, and mudflat environments. Fossils of foraminifera, brachiopods, crinoids, corals, bryozoans are present.

The Death Valley region was still beneath a shallow transgressing and regressing sea; however, in the intervening period since the Eureka Quartzite had been deposited (past ~155 MY encompassing the Silurian and Devonian Periods), Earth had once again seen major changes. The continental plates were converging to form a new super continent, Pangea. The tectonic plate that the Death Valley region was on had drifted up to the equator. The Antler Orogeny had started along the west coast of North America with mountains forming in central Nevada. The oxygen levels were higher than today, spurring plant and animal growth.

In the seas, jawless fish had developed body armor, the first fish with jaws (e.g. sharks and placoderms) appeared, ray-finned fish evolved, and lobe-finned fish developed. The placoderms included the huge, armored predator *Dunkleosteus*. Of all these myriad fish, only the sharks, ray-finned fish (99% of today's fish), and lobe-finned fish survived to the present.

On the distant continental areas, plants and animals had invaded. Vascular (containing tubes for transporting water and nutrients) plants had developed further. Lycopsids (like living club mosses) reached 100 feet high and dominated the Pennsylvania coal swamps, sphenopsids (like today's horsetail), true ferns (abundant today) and seed bearing ferns (would later develop into cycads) became 30 feet tall. This green, forested landscape had been invaded by the Late Devonian by arthropods (millipedes, scorpions, mites, spiders, and insects). Amphibians had struggled ashore and reptiles had laid the first eggs. The land was alive.

Mesozoic Era (251-66 MY)

Above the Sea, Age of the Dinosaurs

16.0 Mesozoic Volcanic Rocks
17.0 Mesozoic Plutonic Rocks

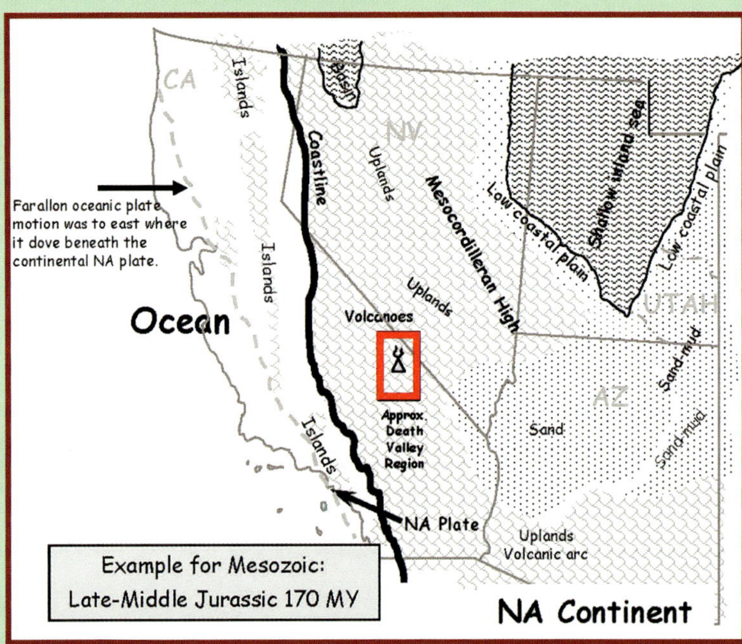

During the Late Paleozoic Era, an active margin had formed with the Farallon oceanic plate diving beneath the continental North American Plate. The result was mountain building, earthquakes, plutons and volcanic activity throughout the Mesozoic Era and into the Cenozoic Era. By the Late-Middle Jurassic (~170 MY), an inland sea had intruded from the north into western Utah and eastern NV. By the Cretaceous, the Western Interior Seaway would bisect the continent. The mountainous Death Valley region's terrain saw erosion of previous sedimentary deposits and new volcanic deposits from within and exterior to its boundaries. Compression forces, also resulting from collision of the two plates, folded and faulted pre-existing rock beds.

Earth's plant and animal populations suffered major mass extinctions at the end of the Triassic and Cretaceous.

Reference, Scotesse, Blakey, Fiero, Prothero, Troxel

Mesozoic Eon (251-66 MY) - Above the Sea, Age of the Dinosaurs

The Mesozoic Eon was a violent timeframe along the west coast of North America, in to the Death Valley region, and on to Nevada and the Southwest. Pangea was disassembling. The west coast's once passive margin of the early and middle Paleozoic Era had been replaced by an active margin due to tectonic plate convergence along that coast. During the early Mesozoic, the Death Valley region was intermittently below a shallow sea. It would remain above the sea after ~200 MY. Four mountain building episodes (Sonoma, Nevadan, Sevier, and Larimide) occupied approximately half of the time from 280 to 40 MY. As a result, the Death Valley region became primarily eroding uplands which did not facilitate the deposition or retention of sedimentary rocks. The only Mesozoic sedimentary rocks in the Death Valley region are found in the Butte Valley Fm. (8,000 feet of Triassic limestone with ammonites, brachiopods, crinoids, and corals which was deposited in the southern Panamint Range) or are found in the Inyo Mountains. Also resulting from tectonic plate convergence were subsurface thrust sheets, volcanoes, plutons (magma domes rising toward but not reaching the surface), and earthquakes extending into and beyond Death Valley. The thrust sheets bent the subsurface beds of the pre-existing Paleozoic marine and Proterozoic rocks into folds.

Hadrosaurs stop for a drink.
Heinrich Harder

The Mesozoic Eon is best known as the "Age of Dinosaurs." Although we have no dinosaur or reptile fossils specifically within the Death Valley region, we can correlate to nearby areas and determine viable inhabitants such as *Coelophysis* (Triassic), *Dilophosaurus* (Jurassic), and *Lambeosaurus* and *Troodon* (Cretaceous). Marine species along the nearby coast to the west and the Western Interior Seaway to the east would have included invertebrate *Ammonites* and large marine reptiles *Shonisaurus* (Triassic-ichthyosaur), *Dolichorhynchops* (Jurassic-plesiosaur), and *Tylosaurus* (Cretaceous-mosasaur). Plants would have included cycads, ferns, conifers, and horsetails plus the Late Jurassic arrival of angiosperms… flowering plants, that soon dominated Earth. Throughout the Mesozoic, temperatures were significantly hotter and sea levels higher than today. The Mesozoic ended 66 MY with a mass extinction possibly induced by an asteroid impact. Non-avian dinosaurs, ammonites, and large marine reptiles were among the casualties.

16.0 Mesozoic Volcanic Rocks

Although we lack sedimentary rocks to help us more adequately define Death Valley's environment, we have the fine grained volcanic rocks in the Avawatz and Inyo Mountains and the Coso, Argus, and Last Chance Ranges. They vary in color depending on their mineral content.

The rocks shown on this page are Triassic metavolcanic (volcanic rock showing evidence of having been subjected to metamorphism) rocks from the northwestern Avawatz Mountains.

This dark red metavolcanic rock has iron and may have been deposited as andesite tending toward basalt.

The two metavolcanic rocks above were likely deposited as rhyolite approaching andesite.

These two metavolcanic rocks were originally deposited as rhyolite.

The wide diversity in colors results from differences in mineral content. The darker rocks contain more iron and magnesium. The lighter color rocks contain the least iron and magnesium and the most silica. They are rhyolite. Additional iron and magnesium leads to andesite and finally the dark, almost black basalt.

17.0 Mesozoic Era (251-66 MY) - Plutonic Rocks

Plutonic rocks, also known as intrusive igneous rocks, are formed when magma rises near to but does not reach the surface. They cool beneath the Earth's surface and form crystals as seen in these photos. The longer it takes the magma to cool, the larger will be the crystals. Chemically, the plutonic rocks granite, diorite, and gabbro are the same as volcanic rocks rhyolite, andesite, and basalt, respectively. Mesozoic plutonic rocks are widely scattered throughout the Death Valley region as seen in **Figure 6**.

The dark crystals seen in the plutonic rocks; starting with the highest iron, magnesium, and calcium content; are the minerals olivine*, pyroxene*, hornblende*, muscovite, and biotite, respectively. The light crystals include quartz and feldspar (orthoclase and plagioclase). The lightest color plutonic rocks contain up to 35% quartz (silica), 45% feldspars, and minor calcium, iron, and magnesium, and are known as granite (upper left hand photograph). Their equivalent in volcanic rocks is rhyolite which comes from the most violent volcanoes (e.g. Krakatoa or Mt. St. Helens) due to the magma's high silica content (high viscosity). The slightly darker plutonic rock is diorite which contains no quartz, added iron and magnesium, and 30-50% feldspar. Its volcanic equivalent is andesite. The darkest plutonic rocks are known as gabbro and contain a predominance of the iron and magnesium minerals with some plagioclase. Its volcanic equivalent is basalt which results from benign volcanoes as found in the Hawaiian Islands. If the plutonic rock is comprised entirely of iron and magnesium minerals (olivine and pyroxene), they are referred to as being ultra mafic. These particular rocks are Triassic Period (251-202 MY) and were located in the northwestern Avawatz Mountains. **contains iron and magnesium*

While the Triassic plutonic rocks shown here were cooling far below the Death Valley region, marine invertebrates like ammonites (top) and reptiles like ichthyosaurs swam in the Triassic sea to the west and preyed on other marine animals.
Heinrich Harder

Cenozoic Era (65.5-2.6 MY)

Savannahs to Basins & Ranges, Age of Mammals

18.0 Titus Canyon Fm.
19.0 Late Miocene to Early Pliocene Volcanic Rocks
20.0 Pliocene Epoch Siltstone
21.0 Pliocene Epoch Borates
22.0 Pliocene to Pleistocene Epoch Gypsum
23.0 Late Pleistocene to Holocene Epoch Fanglomerates
24.0 Holocene Epoch

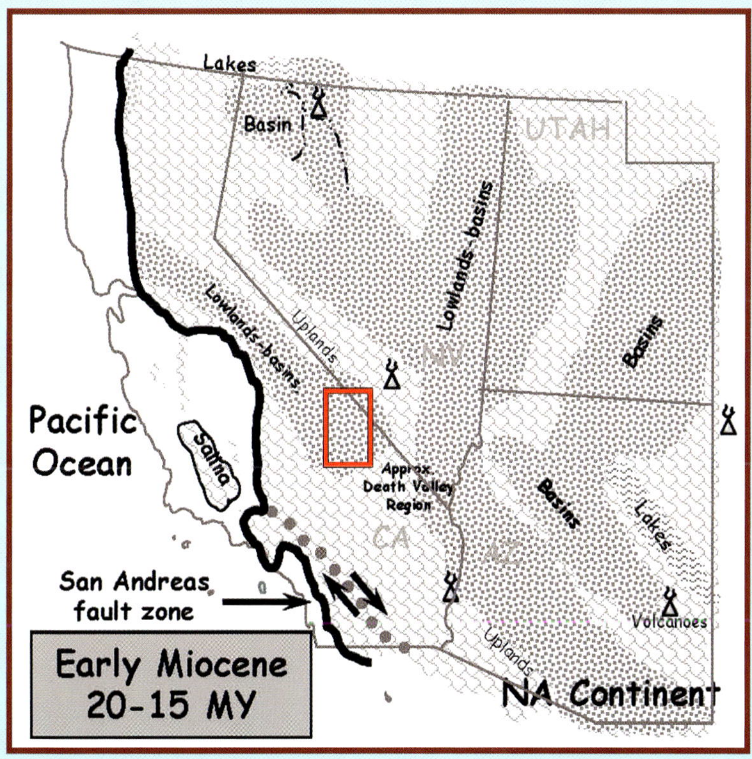

During the Miocene, the sea level was significantly both above and below today's level. The Late Miocene global climate was much as today. During the early Miocene (20-15 MY), the Death Valley region was partially in flatlands, with volcanism widespread in the continental interior. Prior to ~15 MY, plate tectonics changed along the coast west of Death Valley. Subduction ceased and a right lateral boundary formed creating the San Andreas Fault zone and other right lateral zones as far inland as Death Valley. Great Basin extension also began to form basins and ranges in Death Valley, Nevada, and surrounding areas.

Reference: Fiero, Scotese, Blakey

Cenozoic Era - Tertiary Period (65.5-2.6 MY)

The Cenozoic Era started with a continuing confrontation of oceanic and continental tectonic plates along the California coastline resulting in Death Valley facing a continuation of volcanic and plutonic activity and a continued folding of the horizontally deposited earlier rock layers. As in the Mesozoic, the Paleocene and early Eocene Epochs were topographic highs that had few sedimentary rocks formed. Subsequently, they were eroded. The growth of the Sierra Nevada accelerated. Sea level fluctuated above and below today's level.

Initially, the Cenozoic Era climate was warm and humid (tropical). The Eocene Epoch was the hottest of the Cenozoic Era. Cooling and drying trends started approximately 20 MY. Cenozoic plants changed from tropical forests to savannahs to grasslands and finally to deserts. It was the Age of Mammals with once diminutive mammals growing to prodigious size with a multiplicity of genera of primates, multituberculates, creodonts, horses, camels, llamas, beardogs, cats, bears, rhinoceros, giant sloths, mastodons, canines, mammoths, etc.

During the middle Miocene Epoch, changing tectonics resulted in the a) an end to subduction, b) creation of right lateral faults in southern California and Death Valley, and c) extension of the Great Basin producing rising mountain ranges and down dropping basins.

18.0 Titus Canyon Formation (~38-34 MY) Late Eocene Epoch

By the Late Eocene Epoch, mountain building had slowed, worldwide temperature had been dropping, and regional aridity had been increasing since the hot lush days of the Mesozoic. By the Middle to Late Eocene Epoch, the Death Valley region had changed from highlands to flat savannahs consisting of grasslands, scattered trees, and drought resistant undergrowth. The savannah was populated with titanotheres, early horse *Mesohippus*, tapirs, deer-like *Leptomeryx*, etc. The area contained lakes and streams, some of which drained from the Sierra Nevada Mountains, which occasionally flooded creating conglomerates which typify the lower Titus Canyon Fm.

An outcropping of Titus Canton Fm. conglomerate displays rounded clasts of differing size and earlier lithologies in a poorly consolidated sandstone matrix.

Three-toed, early horse *Mesohippus* ran across Death Valley savanna. It was the size of German Shepard dog (2' at shoulder) and browsed on twigs and fruits.
Heinrich Harder

Titanothere (*Brontotherium platyceras)* browsed on late Eocene Epoch Death Valley savanna environment. The savanna might have looked like this picture except for some additional shrubs and trees for the titanotheres and other browsers.
Charles Knight

Repetitive floodplain events formed Titus Canyon Fm. Note the different layers of large, rounded clasts followed by a layer of smaller clasts, and a layer of fine clasts, etc., etc.

19.0 Late Miocene to Early Pliocene Epochs Volcanic Rocks

[Late Late Miocene- Pliocene (~12-5 MY)]

There are numerous examples of Late Miocene and Pliocene Epoch volcanics in the Death Valley Region. Prior to approximately 15 MY and extending back to Late Paleozoic times, volcanic activity was caused by tectonic plates colliding with the west coast and forming subduction zones. By 15 MY, tectonics had changed and plate motion west of Death Valley was lateral (e.g. San Andreas Fault). In the Death Valley region, fault zones such as Furnace Creek, Death Valley, Garlock, etc began forming. Also, Great Basin extension (pulling apart in roughly east-west direction) began. Volcanic activity in the Death Valley region continued with rhyolitic to basaltic flows and rhyolitic ash falls covering parts of the Funeral, Grapevine, Cottonwood, Black, and Avawatz Mountains; Alexander Hills; and Panamint, Argus, and Coso Ranges.

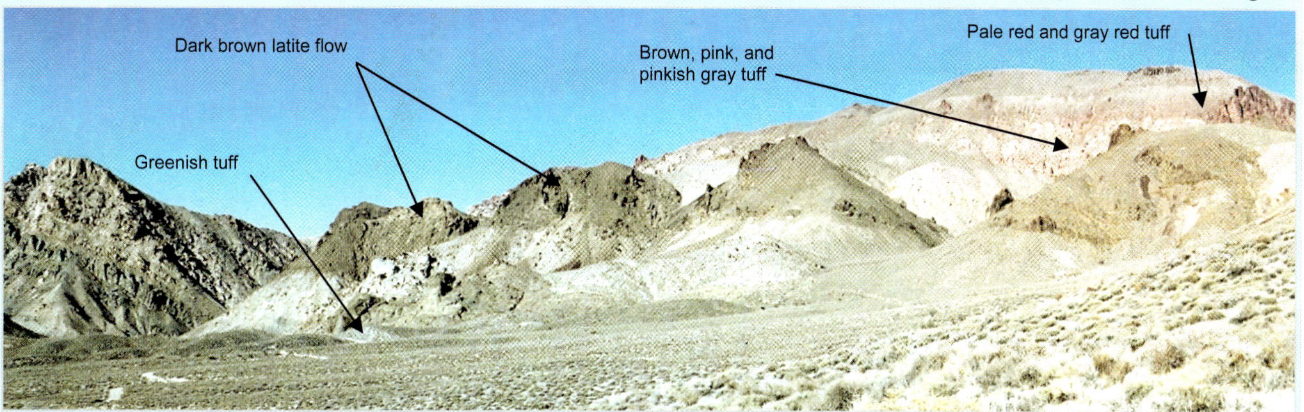

In the Greenwater Range, the Titanothere and Titus Canyon area abounds with colorful volcanic debris. Above, ash falls and flows along with a latite flow and provide brown , tan , green, pink, red, and gray colors. The Timber Mountain Caldera was the source of the different colored sequences of the 12-6.5 MY ash.

Titus Canyon andesitic (left) and rhyolitic (right) ash fall tuffs supplied by the Timber Mountain Caldera located near Yucca Mountain, NV.

A roadcut near Shoshone, California furnished the 11 MY sample of welded tuff to the left. Other colors of welded tuff (fragmented volcanic rocks explosively ejected in flows where heat and overburden produce a welded zone with flattened shards.), at this roadcut are tan, brown, yellow, and white. The black streak to the right is a dense black vitrophyre (glassy) zone that is bounded by brown vitrophyre.

19.0 Late Miocene to Early Pliocene Epoch Volcanic Rocks (continued)

[Late Late Miocene- Pliocene (~12-5 MY)]

In the southern Greenwater Range, near Salsberry Pass (west of Shoshone on Highway 178), a number of volcanic events can be seen. They are the Rhodes Tuff (~10-8 MY), Sheephead Andesite (~8-6MY), Shoshone Volcanics (5-6 MY), and the Greenwater Volcanics (5 MY). Events include lava flows, ash fall tuffs, ash flow tuffs, and welded tuffs. The landscape contains yellows, pinks, and grays.

The Rhodes Tuff upper unit is poorly consolidated, pink, rhyolitic tuff. The middle is rhyolitic ash-flow tuff welded in various shades of red and brown as seen in the accompanying photograph. The lower unit is rhyolitic air-fall tuff, yellow, and layered to massive. Overall, it is up to 1400 feet thick. At the bottom of the Rhodes Tuff, there is a 3 to 6 foot thick layer of obsidian.

Sheephead Andesite can be described as dark lavender andesite, commonly with large white phenocrysts of plagioclase occurring in discontinuous flows occurring in the southern Greenwater Range and southern Black Mountains. Its primary origin is andesitic lava flows. It overlies the Rhodes Tuff.

The Shoshone Volcanics (5-6 MY) are found in the Dublin Hills, Greenwater Range, and southern Black Mountains. Starting from Salsberry Pass and continuing northerly, the volcanics are seen for approximately 20 miles. Thickness is up to 2,700 feet. They are primarily rhyolitic lava flows and ash fall tuffs in repeating color sequences of yellow, gray, pink, gray, and yellow. Pink and gray forms cliffs while yellow forms benches. In their southern habitat, ash flow tuffs are on the bottom of the aforementioned sequence. The gray cliffs are a vitrophyre (glassy volcanic) while the pink and yellow are crystalline. A portion of each glassy gray layer (photographs lower right) is of perlitic structure where cooling resulted in cracks forming small concentric circles or spheroids. Some have hollows or mini caverns that contain quartz crystals, agates, or opals. Go to Salsberry Pass and look north and east.

Welded tuff from the middle of the Rhodes Tuff has a glassy rhyolite-andesite matrix. The flattened shards are the same shape as the Shoshone roadcut's welded tuff.

Sheephead Andesite, below and east of Salsberry Pass, Greenwater Range.

Round, reddish balls or spheroids formed in this gray, glassy matrix. Where this vitrophyre is exposed, often the spheres weather out and land at the base of the exposure. Photo directly below shows small "caverns" also found in the perlitic rock.

Shoshone Volcanics' ash flow tuff.

The air-fall tuffs and lava flows of the early Pliocene (5 MY) Greenwater Volcanics (above) overlie the eroded surface of the Shoshone Volcanics. They are found in Greenwater Valley, southern Greenwater Range. These volcanics mark the end of rhyolitic volcanics in the area. Diminishing basaltic volcanic activity would continue into the Late Pleistocene.

The above basalt sample, is dark due to a much higher iron and magnesium content than the rhyolitic and andesitic volcanics we have previously discussed.

Up until now we have discussed rhyolitic and andesitic rocks which were birthed by high viscosity, violent volcanoes. There are numerous locations throughout the Death Valley region where the low viscosity, almost benign volcanoes poured forth flows of Late Miocene to Late Pleistocene basaltic lava. Resting Spring Fm. (14.1 MY) in the Alexander Hills, Epaulet Peak (7.5 MY) in the southern Black Mountains, Lemoigne Canyon basalt (6.2 MY) in the Cottonwood Mountains, and basalt flows in the Coso and Argus Ranges are examples.

Basalt boulders are strewn across this plateau in the Greenwater Range. Colorful Artist Drive Fm. in the Black Mountains is in the background.

During the Late Miocene to Early Pliocene, the continued trend toward cooling and aridity changed the once short grass and scrub environment (browse) to tall grasslands. Herbivores needed to adapt their teeth to the high silica content grasses. Those that didn't became extinct. Indeed, the Late Miocene saw the largest extinction of North American land mammals in the Cenozoic. Yet, there was a wide diversity of life during Late Miocene to Pliocene time interval which likely frequented the Death Valley region and its lakes. Fossils of animals found in the Death Valley region or nearby locales in California and Nevada that lived in the Late Miocene and transitioned into the Pliocene include: *Pliohippus* an early horse; *Gomphotherium* a shovel tusker, early elephant; *Megalonyx* a ground sloth the size of an ox; *Pseudaclurus* early cat the size of a lynx; *Borophagus* a 45 pound bone crushing dog; *Platygonus* an early peccary, and *Teleoceras* a grazing rhino seen to the right.

Teleoceras was a grazer that developed high crown teeth for grasses. It was the most abundant North American, Miocene-Pliocene rhino. It may have grazed along waterways. Its fossils were found in Titus Canyon, NE, NM, FL, Mexico, TX, CA, NV, OK, KS. *R. Bruce Horsfall, 1912*

20.0 Pliocene Epoch Siltstone [(~5 MY)]

With continued Great Basin extension established in the Miocene Epoch (~15 MY), basins were dropping and ranges rising. The climate was wetter than today and lakes had formed. The lakes provided water for animals and collected minerals for later evaporites. The Pliocene Epoch generally followed the Miocene Epoch's cooling trend with glaciers in the California Sierra Nevada by approximately 3 MY. Grasslands supported those animals that successfully adapted (developing hypsodont teeth) to eating high silica content grasses in lieu of softer browse. Today, Pliocene siltstones provide a record of animals that frequented the lakes.

Ripple marks and mastodon tracks, Pliocene siltstone, Black Mtns.

Camel tracks in Pliocene siltstone, Black Mountains.

Horse tracks in Pliocene siltstone, Black Mountains.

Pliocene animals that would have frequented these lakes include *Dinohippus* (likely ancestor to modern horse), *Megatylops* (extinct 2,200 pound camel), different species of mastodon, *Canis edwardii* (extinct wolf-coyote), *Osteoborus* (extinct "bone crushing dog"), *Felis concolor* (extant mountain lion), *Lynx rufus* (extant bobcat), *Meganteron* (extinct, 275 pound, dirk toothed cat, *Arctodus simus* (extinct, giant short faced bear), and a host of others.

Mastodons (8' tall at shoulder) inhabited the Death Valley region in the Pliocene and Pleistocene. *Bob Giuliani*

21.0 Pliocene Epoch Borates [(~5 MY)]

Borates are minerals that contain the element boron. There are 238 known borates (~31 in Death Valley). Anhydrous borates form in igneous or metamorphic rocks in the absence of water. Hydrous borates develop in sedimentary rocks at the bottom of playas in arid regions. The hydrous borates are white, colorless, or transparent and are brittle and relatively soft. Many of these hydrous borates are soluble in water which helps to explain why concentrations are found at the bottom of playas. They are referred to as evaporites. In Death Valley National Park, the borates colemanite, probertite, and ulexite are found in abundance in the lower Furnace Creek Formation in the region bounded by the Furnace Creek, Death Valley, and Sheephead Fault Zones; with Harmony Borax Works on the northern extremity; and Amargosa Borax Works on the southern extremity. There are traces of other borates such as borax, tincalconite, nobelite, gowerite, bakerite, inyoite, and meyerhofferite. There are approximately twenty-three borate mines in that area. Deposits are located near fractures that provided conduits for volcanic materials and borate enriched fluids to reach the surface. Pliocene Epoch (6 MY) spring discharge, driven by regional volcanic activity which, in turn, was driven by Great Basin extension, seems to be responsible for the borate deposits. Borates are also found in the Saline Valley and Searles Lake.

This three foot long borate rock trail sample, from an open pit mine, is comprised of colemanite and probertite. A close-up of the colemanite present in the sample is seen to the lower left photo. To the upper right is a close-up of a probertite crystal from the sample. A close-up of colemanite crystals from a different sample is seen on the bottom center photo.

A close-up of a radial probertite crystal from the three foot sample to the left.

Close-up of massive colemanite crystals from three foot sample above.

Close-up of prismatic colemanite crystals.

Satiny, white, ulexite from Boron mine appears fibrous unlike "cotton balls."

Borates were discovered in 1873 (Slate Range by Westerville, et. al. and in Death Valley by unknown prospectors), 1874 (Searles Lake by Searles and in Saline Valley), and 1881 (Death Valley Winters and Daunet discovered ulexite). Colemanite was first discovered in 1882 at Death Valley's Monte Blanco mine in the northern Black Mountains. It was named after William Tell Coleman, owner of U. S. Borax. Eventually, colemanite was also found at the Corkscrew mine (Black Mountains), Lila C. and Billie mines (Greenwater Range), in mines at Yermo and Boron, California, and in Nevada. "Cottonball" ulexite is the borate that was discovered in 1882 by Winters and Daunet and eventually processed at Harmony (1883-1888), Amargosa Borax Works (1884-1888), and Eagle Borax Works (1882-1884), respectively. Also in the environs of Death Valley National Park is Saline Valley's Conn-Trudo Borax Works where cottonball ulexite deposits were discovered in 1874. Production did not start there until 1889. Ulexite is often found in conjunction with the colemanite and probertite deposits. Death Valley mines that mined probertite include the Billie, Gower Gulch, Corkscrew, Widow 3 and 7, and Played Out. Borate mining ceased in Death Valley National Park in 1927.

While there are traces of many borates in Death Valley, the borates that were mined in the Death Valley region's northern Black Mountains, Greenwater Range and Furnace Creek playa were colemanite (hydrous calcium borate, $Ca_2B_6O_{11} \bullet 5H_2O$), ulexite (hydrated sodium calcium borate, aka "TV Rock," $NaCaB_5O_9 \bullet 8H_2O$), and probertite (a lower hydrate of sodium calcium borate, $NaCaB_5O_9 \bullet 5H_2O$). As a revelation to many, borax (hydrous sodium borate, aka tincal, $Na_2B_4O_7 \bullet 10H_2O$) was not mined in Death Valley National Park. Ulexite was mined at Harmony and Eagle Borax Works. Onsite processing purified and converted the ore to borax (tincal) with some tincalconite. The closest borax mining area is Searles Lake which started operation in 1874. Borax samples at Searles, or elsewhere, eventually dry out and loose five water molecules and become the dry, white, chalky looking tincalconite ($Na_2B_4O_7 \bullet 5H_2O$).

Coleman's Harmony Borax Works (1883-1888) near Furnace Creek gathered and processed cottonball ulexite. 20 Mule Teams hauled it to a railhead at Mojave. Barn is behind wagons. Settling vats are behind mules. Boiler is farther back. *Death Valley National Park (U. S. Borax Collection)*

Daunet's Eagle Borax Works (1882-1884) in west central Death Valley was the first of the area's processing plants. Financial difficulties and a wife drove Daunet to suicide. *Death Valley National Park (U. S. Borax Collection)*

Amargosa Borax Works (1883-1888) was over 1,500 feet higher elevation than Harmony Borax Works. In the hot summers when ulexite could not be processed at Harmony, it was shipped to Amargosa for processing and later shipped to Mojave. Borates were not "mined" at Amargosa. Processing facilities were to the right (north) in the photo. *Death Valley National Park (U. S. Borax Collection)*

Conn-Trudo Borax Works (1889) , Saline Valley, Death Valley National Park. *Death Valley National Park*

Greenwater Range borate town "New" Ryan (1915-1927) replaced "Old" Ryan (1907-1915) which was at Lila C. *Death Valley National Park (U. S. Borax Collection)*

22.0 Pliocene to Pleistocene Epoch Gypsum [(6-1.7 MY)]

Gypsum (hydrous calcium sulfate, $CaSO_4 \bullet 2H_2O$) is soft, flexible, colorless, and has flat crystals. It is an evaporite. It originates by evaporation of saline lakes in an arid or semi arid environment. Gypsum is a relatively common mineral and can be found throughout the world. In the Death Valley region, gypsum can be found in the northeastern Avawatz, northern Black and Owlshead Mountains; western Panamint and Coso Ranges; and the Confidence Hills. The gypsum from the Furnace Creek Fm. has been dated approximately 6 MY while that in the Confidence Hills Fm. has been dated at 1.7-2.0 MY. It formed by evaporation of the Pliocene and Pleistocene lakes that were once in the Death Valley region. In the northeast edge of the Avawatz, gypsum bearing strata is 150 to 400 feet thick with outcrops occurring in a 1 by 9 mile area. The strata is greenish or reddish (see below left) and comprised of sands and clays with large quantities of gypsum. Pure gypsum, selenite, (see below right) is found in the China Ranch area.

This sample was collected in the north east corner of the Avawatz Mountains. The greenish-gray and reddish beds are sand, clay and gypsum.

This gypsum (selenite) sample is white, transparent, and cleaves into flat layers as seen in this 4" x 5" sample from the China Ranch area.

Nancy Lengner beside twisted beds of reddish gypsum bearing strata in a canyon in the Avawatz Mountains, near the Garlock fault zone.

In 1912, William G. Kerckhoff bought Whispering Kelly's Saratoga Springs rock-salt property and several square miles of gypsum property in the valley to the south of Saratoga Springs and organized the Avawatz Salt and Gypsum Company which held numerous claims. In 1913, a 10 mile spur from the T&TRR to the deposits was discussed. Kerckhoff found out that shipping costs would make any ventures unprofitable; however, he still held on to the properties until his death in 1929.

In 1915, Acme Cement & Plaster Co. opened the Gypsy Queen Group of four gypsum mines near China Ranch. The T&TRR built a 1.3-mile spur from Morrison (aka Acme) siding in Amargosa Canyon (five miles south of Tecopa) up Willow Creek and past China Ranch to the site of Acme. It produced about $100,000 from 1915 to 1918 but was closed in 1918 when a large cave-in killed two men. The gypsum was in 1" to 2" stringers in clay.

One of the approximately half-dozen adits of the Gypsy Queen Mine. The sign reads, in English and Spanish, "DANGER!, Extremely Loose Soil, DO NOT ENTER." Chain link fences discourage entrance. Another adit's sign warns of the cave-in that killed two men. Flat fragments of gypsum abound in the area.

23.0 Late Pleistocene- to Holocene Epoch Fanglomerates

[(~.300MY - Present)]

The Death Valley region has numerous alluvial fans created by rain waters draining through a narrow canyon onto the "valley" floor. Viewed from above, the alluvial fan has the shape of an open hand fan. As the water and debris issues forth from the mouth of the canyon, large boulders are deposited near the mouth and center of the fan. As the energy dissipates to the fan sides and downhill, the size of the debris changes to cobbles, gravel and eventually sand and mud. The composition of the fan is based on the lithology of the rocks in the canyon from which they were eroded. An excellent photo of an alluvial fan is seen below. Although there are exceptions, most alluvial fans are late Pleistocene to Holocene in age.

A fanglomerate is a sedimentary rock consisting of dissimilar materials (conglomerates or breccias of various lithologies) that were originally deposited in an alluvial fan and since have been cemented into solid rock. The cement of choice in the Death Valley region is called caliche. Its source is the calcium carbonate laden waters that wash the debris down onto the fan.

This excellent photo of the alluvial fan near Badwater shows its shape, handle, and braided streams. *Marli Bryant Miller*

Two fanglomerate samples. To the near right is a conglomerate with varying sizes and lithologies. The far right, is a breccia with a fairly consistent lithology. Both are well cemented and do not crumble to the touch.

Directly above, is a fanglomerate composed of breccia with large clasts (one with a manganese fern) in a red (iron oxide) matrix. To the right, is a breccia with small clasts of differing lithologies.

During the Pleistocene, Death Valley was generally wetter and cooler than today. Lakes and playas were present. Three drainage systems (from Mono Lake and along a series of additional lakes farther south of Mono, from Mojave River, and from Amargosa River [post 186 KY]) intermittently fed the Death Valley region and formed lakes like Manly (600' deep) and Panamint (900' deep). Although there may have been up to 20 glacial periods during the last 2 MY, Death Valley had no glaciers.

Lake Tecopa, fed by the Amargosa River, deposits include fossils of 5-species of camels, pronghorn antelope, small and large horses, mastodons, and mammoths. The extant kit fox, gray fox, black bear, bison, bobcat, lynx, and mountain lion along with the extinct dire wolf, Smilodon, American lion, and Columbian mammoth likely were in the Death Valley region. Approximately 10 KY many large mammals became extinct in North America.

This 3' by 3' fanglomerate was found in a small canyon in the Funeral Mountains. It has small clasts of varying lithologies that are both rounded (conglomerate) and angular (breccia). The clasts are cemented by caliche; however, the bond is not good enough to prevent some clasts from separating and leaving holes on the fanglomerate surface.

Smilodon (aka saber toothed cat or saber-toothed tiger) was an ambush predator of ground sloths, bison, deer, camels, horses and young mammoths. It was powerfully built and had a short tail. The largest species had 11' scimitar canines and weighed over 700 pounds. They became extinct around the beginning of the Holocene. *Charles R. Knight*

Columbian mammoth (aka Imperial mammoth) was ~12-13' at the shoulder, had 16' curved tusks and grazed on grasses. It ate 700-800 pounds of food per day. It became extinct in the late Late Pleistocene (~10 KY).

R. Bruce Horsfall 1912

24.0 Holocene Epoch

(10 KY - Present)

Great Basin extension continues to the present with ranges being uplifted and basins being down-dropped. Plate motions along the coast still transmit a small right lateral shear requirement to the Death Valley region; however, earthquakes are extremely rare. The Ubehebe craters were formed approximately 6 KY by steam explosions which created minimal volcanic debris. Wind and water continues to weather and erode the uplifting ranges creating material for new sedimentary rocks and filling basin floors. Winds and topography have created over six major sand dunes.

At the beginning of the Holocene (10 KY), the Death Valley region was wetter than today. Increased aridity, in part due to the uplifted Sierra Nevada and Transverse Ranges, began approximately 6 KY. Transition to today's desert vegetation was completed in the 6-3 KY time span. The specific time of transition was a function of altitude. As the climate became more arid, plants found to the south "migrated" to the north, plants "migrated" upward (3,500 ft?) in altitude, and plant populations were altered.

Gone are the huge mammals like mastodons and mammoths and other diverse animal populations that once roamed the Death Valley region. With the exception of a few elk or deer that frequent the north western area, Desert Big Horn sheep are the largest herbivore. The Mountain Lion (*Felis concolor*) is the largest carnivore. Man arrived and made his presence known in the Holocene.

Wind and water shaped the topography near Zabriskie Point.

Devil's Golf Course, silts and salts from Holocene Lake Manly.

Wind blown sand created this 2" ventrifact, Ibex Hills.

Wind formed horizontal salt crystals, Badwater area.

Native American petroglyphs, Funeral Mountains.

2" piece of slag from smelter in Nopah Range.

a few more of The Magnificent Rocks of Death Valley

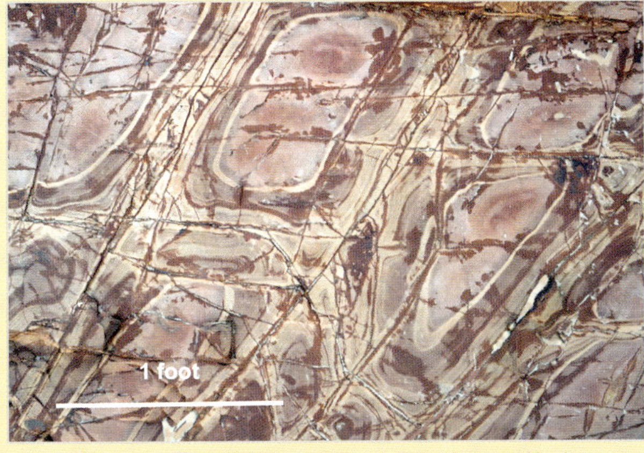

Liesegang Rings, Kingston Peak Fm., Sperry Wash.

Onyx (Chalcedony, SiO_2), Argus Range.

Colonial coral, Bird Spring Fm., southern Death Valley.

Lead ore and yellow sulfur in Noonday Dolomite. Red from hematite.

Garnets in mudstone, Kingston Range.

Artist Drive Fm., northern Greenwater Range and Black Mountains.

Mudstone (2" X 6"), Ibex Fm., Saddle Peak Hills.

Fool's Gold (FeS) and calcite ($CaCO_3$) in iron ore, Kingston Range.

Tincalconite (white) on pyramidal hanksite crystal, Searles Lake.

Travertine ($CaCO_3$), Funeral and Grapevine Mountains.

One inch thick vein, amethyst (purple quartz, SiO_2), Bullfrog Hills.

Erosion, balancing acts (2), Sperry Wash.

Salty silt or silty salt… which is it? Avawatz Mountains.

Sourdough Limestone ($CaCO_3$), western Panamint Range.

One inch thick, pink calcite in mudstone, Kingston Range.

One inch thick, yellow-brown calcite, Funeral Mountains.

Four inch sample, trona ($Na_3(CO_3)(HCO_3) \bullet 2H_2O$), Searles Lake.

Fourteen inch sample, Jasper (Chalcedony) Greenwater Range.

Close-up, silver (Ag) in Bonanza King Fm., Greenwater Mountains.

Five inch sample, garnet schist, Funeral Mountains.

Close-up “pink” granite, Avawatz Mountains.

Ripple marks, post 1.7 BY Crystal Spring Fm. mudstone.

Fluorspar (Fluorite, CaF_2), Last Chance Range.

Volcanic trachyandesite close-up, Sperry Wash.

Two inch sample, crinoid (sea lilies) stems, Bird Spring Fm.

Orbicular mudstone (slightly metamorphosed), Kingston Range.

Copper stains on Noonday Dolomite, southern Nopah Range.

Two inch sample, aragonite ($CaCO_3$), Bare Mountain.

Igneous, metamorphic, and sedimentary rocks in a wash. Kingston Peak Fm. and diabase background. S. Black Mountains.

Three inch sample, fine bedded siltstone, Johnnie Fm., Sperry Hills.

Ordovician gastropod (1.5 inches), Ely Springs.

Coral and gastropod fossils, Bird Spring Fm.

"Dog Tooth" Calcite, Bare Mountain. 1 inch "teeth."

"Manganese Fern," Pyrolusite (MnO_2) on sandstone.

Calcite and Limestone/Dolomite Jigsaw Puzzle, Titus Canyon.

Freshly fractured marble (metamorphosed limestone) sparkles in sunlight, Kingston Wash.

Trilobite "hash" in siltstone, Carrara Fm . Grapevine Mountains.

"Amargosa Chaos" Black Mountains.

Green epidote [$Ca_2Al_2FeOSiO_4Si_2O_7(OH)$] in granite, Avawatz Mt.

Hexagonal, cross striated quartz (SiO_2) crystals in rhyolite, Kingston Range.

Fan gravel and blowing sand at Mesquite Dunes, Stovepipe Wells.

Nine 1”-2” samples of vesicular lava from the same area. Color differences indicate different minerals/events, south Death Valley.

Half inch cubic halite (NaCl) crystals, Searles Lake.

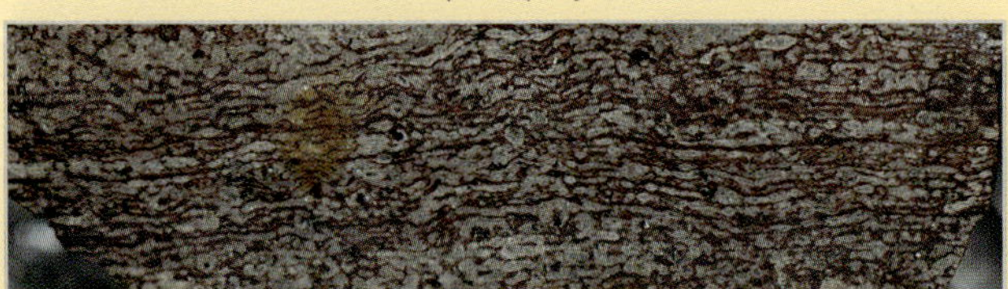

Thin-bedded rhyolite, Dublin Hills, southern Death Valley region.

Granite with unusually large, pink orthoclase crystals, smaller white plagioclase crystals, and smaller black hornblende and biotite crystals, southern Black Mountains.

Greenish and black diabase, two upwelling events, Nopah Range.

Rapakivi granite, Smith Mountain pluton, Black Mountains.

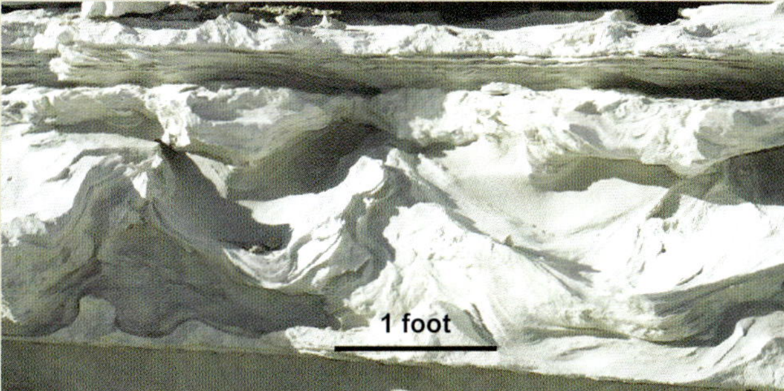

Lava Creek B Ash, soft sediment deformation, Dublin Hills.

Titus Canyon Fm. 5 inch clast, fractured and welded with calcite.

Basin salt (l) and crystalline basement rock (r), Dante's View.

Sulfur (S) ore, Last Chance Range.

Zabriskie Quartzite weathered by Mojave River, Salt Spring Hills.

From Red Pass, Bonanza King Fm. and Miocene tuff.

"Artist Palette." This famed tourist stop displays the multicolored volcanics of the Miocene Artist Drive Fm. Colors come from minerals such as iron (red, pink, yellow), mica (green), and magnesium (purple). Northern Black Mountains.

Spring in salty, muddy, lakebed deposits in Funeral Formation. Furnace Creek.

Copper Canyon turtleback's Precambrian rock is reflected in a seasonal pool bordered by evaporites. Other two turtlebacks are at Badwater and Mormon Point .

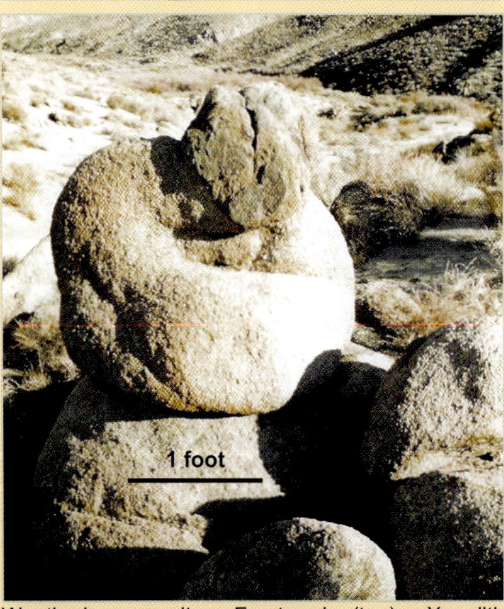

Weathering granite. Fractured (top). Xenolith perches atop "doughnut" (above). Butte Valley.

Limestone and dolomite beds of the Late Pennsylvanian-Early Permian Bird Spring Fm. are tilted 90 degrees and jut above landscape. Striped Butte, Butte Valley.

Appendix

Death Valley Region

Stratigraphic Column

Avawatz Mountains to Last Chance Range

&

Kingston Range to Coso Range

Compiled by: Kenneth E. Lengner

Introduction

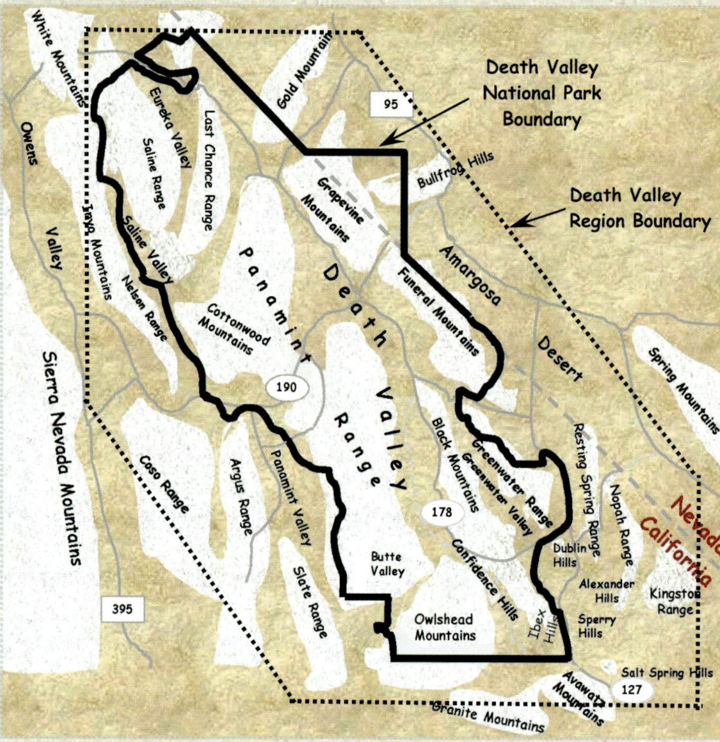

Death Valley region boundaries and mountains.

The following stratigraphic column for the Death Valley region was compiled from multiple sources with the most significant noted in the References. Herein, the Death Valley region is defined such that the Inyo Mountains and Coso Range are the approximate western boundary and the Kingston Range the eastern. Northern boundary is the northern extremity of Death Valley National Park and the Last Chance Range and the southern is the Avawatz Mountains as illustrated by the dashed lines in the accompanying figure. The Spring Mountains or Sierra Nevada Mountains. are not included. The selected boundaries are premised on geological, paleontological, economical, political, and historical considerations. Death Valley National Park's boundaries, to a large extent, are political and economical.

During an estimated 2.5 billion years, the Death Valley region was torn apart, was submerged beneath the seas, was crushed by onrushing tectonic plates, and is currently being torn apart yet again. It witnessed rising and falling sea levels, a wide range of climates, mountain building and erosion, earthquakes, volcanoes, the birth and diversification of life in ancient seas, and the last 250 million years of evolution of terrestrial plant and animal life. It was part of other ancient supercontinents before today's North American continent. This diversity of environments has produced the mountains and fault zones shown in the accompanying figures and sedimentary, igneous, and metamorphic rocks addressed in the following stratigraphic table.

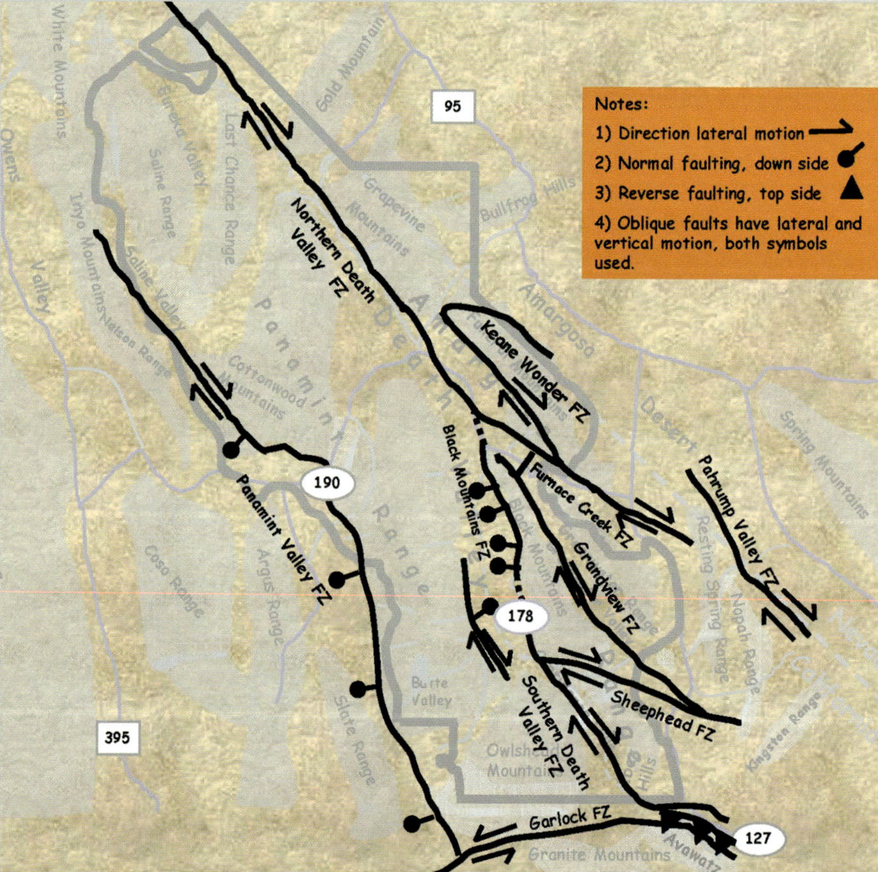

The stratigraphic column starts at the present and continues to the earliest known Death Valley region rocks. In the words of the geologists, it starts at the "Latest" (or Upper) and progresses to the "Earliest" (or Lowest). Because the region covers such a large area, certain rocks or formations may have been deposited in one area while, at the same time, different formations were deposited in another area. In the following stratagraphic column, these "approximately contemporary," but different formations, are identified by like asterisks, coloring, and shading. To provide the reader with perspective, additional information on major events, climate, and plant and animal life is included in the stratigraphic column.

Comments, corrections, updates are solicited.

DEATH VALLEY (DV) REGION STRATIGRAPHIC COLUMN

AGE	FORMATIONS	DESCRIPTIONS	LOCATION
CENOZOIC ERA (0-66MY) Recent Life or Age of Mammals			
QUATERNARY PERIOD (2.6-0 MY) **Holocene Epoch** (0-10 kya)	Surface Deposits: Sedimentary deposits erode down canyons, on to alluvial fans, and continue filling basins. Evaporation. Winds. Rains. Temp.	Fanglomerates, boulders, cobbles, and finer-grained material of various lithology from alluvial fans, landslides, weathering. Desert pavement. Evaporates such as salts (NaCl, KCl). Aeolian sand dunes. Ubehebe steam gas explosion .3 KY (??) Man Arrived in DV. Continued extinction of NA large mammals. Desert climate formed.	Death Valley Region
Pleistocene Epoch (2.6-.01 MY) "Ice Age" Peak of last glacial period 18 KY. 15 glacial cycles in last 1 MY. Glaciers in California Mountains Dire wolves, mammoths, mastodons, giant sloths, saber-toothed cats (Smilodon), horses, camels, long horned bison.	Surface Deposits from: erosion, lakes, and springs	Manly (10-35 KY, 120-180 KY, .620 KY, .8-1.2 MY), Panamint, Saline, Searles, Wingate Pass and Tecopa Lake (2.1-.155 MY) deposits. Sand, silt, marl, claystone; cobble and gravel deposits as bars and embankments; caliche; freshwater limestone. Evaporates: [(borates, gypsum, anhydrite) and (salts: carbonate, sulfate, chloride)] formed as lakes dried up: Travertine deposits formed at springs. Stream deposits. Late Pleistocene Fans: Central DV (~.120 MY), Mormon Point (120,180 MY), Hanapaugh (>.300 MY)	Death Valley Region
	Surface Deposits from: Volcanics from Death Valley region and *elsewhere.	(see table below)	
	Mormon Pt. Fm.- (.18-1.0 MY)	Interbedded mudstones, conglomerates, and tephra beds. Lava Creek B, Bishop Tuff, Upper Glass Mountain.	Mormon Pt., Natural Bridge
	Funeral Fm.	See Below	Funeral Mtns.

Dibekulewe Ash (~.5 MY)	Mormon Point	Blind Springs Valley Tuff (2.2 MY)	Confidence Hills
Lava Creek B Ash (.62 MY)*	Kit Fox Hills	Upper Glass Mtn. Ash (.8-1.2 MY)*	Mormon Point
Cinder Cone Basalt (.69 MY)	Ashford Mill	Middle Glass Mtn. Ash (1.5 MY)*	Furnace Creek
Bishop Ash (.77 MY)*	Mormon Point	Lower Glass Mtn. Ash (1.7-1.9 MY)*	Artist Drive
Shoreline Butte Basalt (1.5 MY)	Ashford Mill	Huckleberry Ridge Ash (2.1 MY)*	Confidence Hills
Confidence Hills Tuff (~2 MY)	Confidence Hills	Emigrant Pass Ash (~2 MY)	Emigrant Pass

AGE	FORMATIONS	DESCRIPTIONS	LOCATION
TERTIARY PERIOD (66.5-2.6 MY) **Pliocene Epoch** ***(5.3-2.6 MY)*** Great Basin extension continues to drop DV basins and uplift/tilt ranges. Glaciers form in California-3 MY. Arid climate in DV. DV Grasslands and some forest. Lakes Felines, canids, horses, camels, llamas, bears, mastodons, antelopes, flamingoes, foxes, rabbits, shrews, rats, mice and birds.	China Ranch Beds	Fanglomerate and siltstone. Camel.	Sperry Hills
	Confidence Hills Fm.- (2.2-1.7 MY)	Sandstone, mudstone, siltstone, halite, gypsum, anhydrite, conglomerate, 3 ash beds, Shoreline Butte Basalt.	Confidence Hills
	Coso Fm.- (~3.3 MY)	Fanglomerate, arkosic sandstone, shale, buff-colored clay; locally iron stained or silicified; minor, thin beds of impure limestone. Contains *pyroclastic* rhyolitic tuff, lapilli tuff, and tuff-breccia, white to light brown, sometimes reddish, moderately well-bedded, often slightly welded, limy, or ferruginous. Horse, llama, bear-dog, mastodon, short-faced bear, rabbit, etc. Conifers, cedars, junipers.	Coso Range
	Funeral Fm.- 1000', (~5.2-1.7 MY)	Basal conglomerate with pebbles, cobbles, and boulders in mud and sand matrix incipiently lithified; gravelly mudstone and sandstone, local travertine and well-lithified conglomerate, large blocks of Paleozoic rocks; olivine-phenocryst basalt; basaltic agglomerate (at explosive vents) consisting of red scoria, bombs, breccia, dikes.	Furnace Creek Wash and Black Mtns.
	Nova Fm.- (5.4-3.1 MY)	Conglomerates, breccias, intercalcated basalt flows.	Cottonwood Mountains, NW of Panamint Mtns.
	Copper Canyon Fm.- 10,000'	Predominantly moderate-red conglomerate. Some intertongues of yellow-gray siltstone and evaporites (e.g. gypsum) in upper portion of Fm. Intercalated greenish-gray andesite and basalt. Freshwater ostracode and gastropod fossils in siltstone. Numerous trackways include bird, camel, horse, mastodon, carnivore, etc.	Black Mtns.
	Furnace Creek Fm.- 5-7000', (~6-<3.5 MY)	Conformably upon Artist Drive. Basal conglomerate contains major borate deposits; predominately lacustrine mudstone and sandstone, some gypsum; granular gypsum and anhydrite; same as second member; light colored mudstone grading into sandstone and conglomerate. Volcanic rocks (e.g. Mesquite Spring, Glass Mountain) throughout formation: fragmented basalt; basaltic flows and intrusions; pale volcanic rocks; vitrophyre breccia.	Northern Black and Funeral Mtns.; Greenwater Range

Ubehebe Hills Basalt (3.7 MY)	Northern DV	Greenwater Volcanics (5 MY) air fall tuffs and lavas. Rhyolitic.	Greenwater Range
Mesquite Spring Tuff Upper (3.1 MY)	Copper Cyn, Furnace Creek	Porphyritic andesite, gray-red ; Andesite flows	Southern Inyo Mtns; Greenwater Range
Mesquite Spring Tuff Lower (3.35 MY)	Kit Fox Hills, Nova Basin	Dark-gray olivine basalt (~4.1 MY); Red, , green, gray basalt (~5.1 MY).	Ash Hills, Towne Pass
Basaltic-andesitic breccia and tuff	E. Argus, Coso Greenwater Range	Curry Canyon Tuff (~3.4-3.6 MY)	Furnace Creek Wash

AGE	FORMATIONS	DESCRIPTION	LOCATIONS
Miocene Epoch (23-5.3 MY) DV topography formed by Great Basin extension and San Andreas right lateral fault which were in-place by ~15 MY. Volcanic activity difficult on plant and animal life. Short grasses and shrubs mixed change to tall grasslands. Migrations. Mixed browsers change to Long legged, grazers (camels, horses, rhinoceroses).	Ubehebe Lakebed Fm.- (<3.2 MY)	Sequence of interbedded conglomerates, tuffs, mudstones, and basalts.	Cottonwood Mtns; Death Valley Wash
	Artist Drive Fm.- 4-5000', (14-6 MY)	Two colorful pyroclastic layers separate three sedimentary layers. Brown weathering mudstone, sandstone, conglomerate; green and buff tuff-breccia; olive or light brown mudstone, sandstone, conglomerate; blue-green, pink, orange tuff-breccia and some basalt; brown conglomerate, green and red volcanic detritus. Felsite and basaltic flows, sills, dikes, and volcanic necks in different parts of formation.	Northern Black Mtns.
	Resting Springs Fm.- 1200', (14.1 MY)	Andesite and basalt flows, sandy limestone, tuff.	Alexander Hills; Pahrump Valley
	Military Camp Fm.- 2800', (15-6 MY)	Clastics and evaporates fining upwards.	N. Avawatz Mtns.
	Bat Mountain Fm.- 1800', (14-17 MY)	Intertonguing conglomerate and sandstone.	Southern. Funeral Mtns.

Miocene Volcanic *tuffs (or ash), basalt, etc.			
Nomlaki Tuff (3.28 MY)*	Artist Drive	Nomlaki Tuff (Lwr)(>3.58 MY)*	Artist Drive
Shoshone Volcanics(6-7 MY)*	S. Black/Grnwter	Sheephead Andesite (8-6 MY)	S. Black/Grnwter
Rhodes Tuff (10-8 MY)*	S. Black/Grnwter	Unconformity Hill (15.7 MY)*	Grapevine Mtns.
Ubehebe Wash (23.7MY)*	N. Death Valley	Redrock Valley (15.0 MY)*	Grapevine Mtns.
Bighorn Gorge (22.7 MY)*	N. Death Valley	Bullfrog (13.3 MY)*	Grapevine Mtns.
Marble Canyon (15.7 MY)*	N. Death Valley	Tiva Canyon (12.7 MY)*	Grapevine Mtns.
Entrance Narrows (12.1 MY)*	N. Death Valley	Rainer Mesa (12.7 MY)*	Grapevine Mtns.
Lemoigne Cyn basalt (6.2MY)	N. Death Valley	Titus/Titanothere Cyn (12-8 MY)*	Grapevine Mtns

Chocolate Sunday Mtn. Pluton, Greenwater Range; Smith Mtn. Pluton, 19 MY, Black Mtns.; Warm Springs Granite, 12 MY, Panamints; Rabbit Hole Springs and Kingston Peak Granite, Kingston Range; etc.

AGE	FORMATIONS	DESCRIPTION	LOCATIONS
Oligocene Epoch (33.9-23. MY)		Volcanoes. DV climate cooling and becoming more arid. Grasses spread.	
Eocene Epoch (55.8-33.9 MY) Late Eocene:	Titus Canyon Fm.- (Late Eocene) –3000'	Sedimentary breccias interfinger with overlying variegated facies which include; siltstone, mudstone, marl, sandstone, conglomerate, limestone, and prevalent conglomerate Crystal lithic tuff, Latite flow. DV is savannah with streams from Sierra Nevada. Some grasslands. Medium to large placentals with Asian migrants. Large diversity mammals. Titanotheres, *Mesohippus, Leptomeryx.*	Grapevine Mtns.
Early, Middle Eocene	None	Sediments lost to erosion. Jungles change to forests.	
Paleocene Epoch (55.8-65.5 MY)	None	Sediments lost to erosion. Strozzi Ranch Rhyolite. Badwater, Copper Cyn. pegmatite dikes. Eroding uplands with hot and humid climate. Small placental, marsupial, and egg laying mammals in tropical jungle.	

MESOZOIC (66-240 MY) Middle Life or - Age of Dinosaurs

West coast tectonics (subduction) resulted in mountain building and plutonic and volcanic events. DV was uplands. The few sediments deposited were eroded. Dinosaurs and other Mesozoic/Paleocene animals roamed North America and DV; however, any *DV fossils were lost to depositional environment or erosion.*

CRETACEOUS PERIOD (145.5-65.5 MY)
First flowering plants. Major land/sea mass extinction at end Cretaceous.

JURASSIC PERIOD (201.6-145.5 MY)
Continent grew west. Pangea breakup.

TRIASSIC PERIOD (251-204 MY)
Near equator.

Granitic (intrusive, plutonic) rocks crop out in the Granite Mtns., Avawatz Mtns., Owlshead Mtns., S. Black Mtns., W. Sperry Hills, S. Quail Mtns., Slate Range, Coso Range, Argus Range, northern Panamint Range, Cottonwood Mtns., Inyo Mtns., White Mtns., and Saline Range. Other Mesozoic, plutonic rocks such as diorite, gabbro, and hornblende gabbro crop out in Black Mtns. *Jurassic: Eureka Valley Pluton, 182 MY, Last Chance Range; Hunter Mountain Pluton, 167-185 MY, Cottonwood Mountains. Cretaceous: Hall Canyon Pluton, Panamints; Skidoo Granite, 67-87 MY, Tucki Mountain; etc.*

Sedimentary and metasedimentary. (Triassic)	Massive and platy fossil bearing limestone and shale.	S. Inyo Mtns.
Metamorphic rocks. (Pre-Cretaceous)	Gneiss, metavolcanic, meta-igneous, and metasedimentary rocks.	Coso Range.
Metavolcanic rocks (Pre-Cretaceous)	a) Metarhyolite, metadacite, and metatuff. b) Diabase and volcanic breccia.	a) Coso Range. b) Last Chance Range.
Metavolcanic rocks. (Jurassic and Triassic)	a) Andesite flows with tuff and breccia. b) Mafic flows, volcanic breccia, tuff.	a) S. Inyo, Avawatz Mtns. b) Argus Range.

Mountain Building
Sonoma Orogeny (280-200 MY).
Nevadan Orogeny (150-140 MY).
Seveir Orogeny (140-80 MY).
Laramide Orogeny (80-40 MY).

Max. packing of continent of Pangea 250 MY.

AGE	FORMATIONS	DESCRIPTION	LOCATIONS
First mammals. First dinosaurs. Major mass extinction end of Triassic.	Butte Valley Fm.- 8000'	**DV emerged from sea-Late Triassic/Early Jurassic.** **Shallow Marine.** Ammonites, brachiopods, crinoids, corrals, etc.	S. Panamint Range (Butte Valley)

PALEOZOIC (251-542 MY) "Ancient Life"

AGE	FORMATIONS	DESCRIPTION	LOCATIONS
PERMIAN PERIOD (299-251 MY) Pangea forming; Sonoma Orogeny (280-200 MY). Island arcs accrete to CA west coast. Protomamals; Worst sea/land mass extinction at end of Permian.	Garlock Series	Tactite, marble, phyllite, schist, hornfels, chert. **Shallow Marine.**	Silurian and Valjean Hills, N. Slate Range, N. Avawatz Mtns.
	Owens Valley Group-2000'	Gray, brown, red, and yellow conglomerate, quartzite, sandstone, siltstone, limestone, and limestone breccias. Brachiopods, ammonoids, conodonts, fusilinids, corals. **Shallow Marine.**	Inyo Mtns.; S. Last Chance, N. Panamint, and Argus Ranges.
	Anvil Spring Fm.-(**Archaic name**)	Limestone, cherty limestone, minor shale, dolomite. **Shallow Marine. (Once referred to Permian to Devonian DV rocks.)** Stripped Butte.	S. Panamint Range (Striped Butte, Butte Valley).
	*Tihvipah Fm.-200'(Late Penn./Early Permian)	Light-gray platy limestone, shaly limestone, medium gray, fine-grained limestone, and calcerous shale. **Shallow Marine.**	N. Panamint Range (Quartz Spring).
CARBONIFEROUS PERIOD Pennsylvanian Epoch (318-299MY) Death Valley still shallow marine and near equator. Glaciers at south pole. Fish, crinoids, ammonites, brachiopods, and sharks. First reptiles.	*Keeler Canyon Fm.-300' (Middle Penn./ Early Permian)	Alternating blue-gray limestone and marble. Brachiopods. **Shallow Marine.** Fusilinids, crinoids, byroad.	S. Last Chance and N. Panamint Ranges; Inyo Mtns, SE Cottonwood Mtns.
	*Bird Spring Fm.-400', (Late Penn\Early Permian)	Pebbly sandstone, coarsely bioclastic limestone, sandy limestone, limestone, dolomite, and chert nodules. **Shallow Marine.** Horn and colonial corals, gastropods, fusilinids, bryozoans, crinoids, echinoids, brachiopods.	Southeastern DVNP; Salt Spring Hills; Nopah and Resting Spring Ranges; Striped Butte; Spring Mtns.
	Resting Spring Shale Fm.- 750'	Olive-gray to olive-brown argillaceous shale and siltstone (Pennsylvanian). **Shallow Marine.** Brachiopods.	N. Panamint Range, Inyo Mtns., Waucoba Mtn, Last Chance Range.
	Lee Flat Limestone-650'	Light gray limestone marble, dark gray limestone, dolomite, chert. **Shallow Marine.**	Argus Range; Inyo Mtns.
CARBONIFEROUS PERIOD Mississippian Epoch (359-318 MY) "Age of Coal Forming Forests," "Age of Ferns," "Age of Amphibians." Antler Orogeny (355-300 MY).	**Chinaman Shale-(Late Mississippian)	Black shale with minor sandstone and limestone. **Shallow Marine.**	Inyo Mtns (Cerro Gordo).
	**Perdido Fm.-350' (Middle/Late Mississippian)	Medium-gray platy limestone, conglomerate, quartzite, sandstone, siltstone, and brown-weathering chert. **Shallow Marine.** Ammonites.	Last Chance, Panamint, and Argus Ranges; Waucoba and Inyo Mountains.
	*Stone Canyon-300' Limestone (Early Miss.)	Abundant black chert beds (<2") interbedded with limestone beds ((2"-4"). **Shallow Marine.**	Warm Spring Cyn; Darwin Hills; Argus Range.
	*Monte Cristo-	Light-gray and dark-gray limestone and dolomite and brown chert. **Shallow Marine.**	Nopah and Resting Spring Ranges.
	*Tin Mt. Limestone-1000' (Early Miss.)	Bluish-gray, fine-grained, cherty limestone. Thin-bedded lower, thick-bedded upper. **Shallow Marine:** offshore bar, lagoons, and mudflats. Foraminifera, brachiopods, crinoids, corals, bryozoans.	S. Last Chance and N. Panamint, and Argus Ranges, Inyo Mtns.
DEVONIAN PERIOD (416-359 MY) First sharks, bony fish, insects, and forests. Oxygen drops and rises. First corrals. Major mass extinction at end of Devonian.	**Lost Burro Fm.-2000' (Middle/Late Devonian)	Lower part: interbedded light- and dark-gray dolomite, quartzite, and sand or cherty dolomite. Nopah Range has sandstone, limestone, and multi-colored breccia. Upper and middle part: light-gray dolomite prominently striped with nearly black limestone and dolomite. **Shallow Marine.** Brachiopods, stromatoporoids, corals, conodonts, gastropods, crinoids.	Last Chance, Panamint, Argus, Nopah, and Resting Spring Ranges; Inyo and Grapevine Mtns.
	**Stewart Valley Fm. *(Not Miocene Fm.. in NV)*	Dark-gray limestone, cherty limestone, sandstone, quartzite, and conglomerate. **Shallow Marine.**	Nopah and Resting Spring Ranges.
	Hidden Valley Dolomite (Early Devonian)	Light-gray to medium-gray, thick-bedded dolomite. **Shallow Marine.** Nearshore. Cup Coral, conodonts.	See Below.
SILURIAN PERIOD (444-416 MY) High oxygen spurs size growth of animals. First air breathing arthropods on land & jawed fish in sea.	Hidden Valley Dolomite Fm.- 300-1,400' (Total Fm., ~2460') (Late Silurian/Early Devonian)	Light–gray to medium-gray, thick-bedded dolomite; nodular chert zone near base. **Shallow Marine.** Rugose corals, brachiopods, conodonts, and crinoids.	Last Chance, Panamint, Argus, Nopah, and Resting Spring Ranges; Inyo, Cottonwood, and Grapevine Mtns.
	Sunday Canyon Fm.-650' (Late Silurian/Early Devonian)	Platy, gray to buff, limy shale/ minor gray limestone interbeds in basal portion, grading upward to thin-bedded blue-gray shaly limestone. **Shallow Marine.** Graptolites.	Waucoba Mountain.

*** or ** with common shading denotes ~ contemporary formations in different locales. (Not in sequence as table appears to indicate.)**

AGE	FORMATIONS	DESCRIPTION	LOCATIONS
ORDOVICIAN PERIOD (488-444MY)	Ely Springs Dolomite-700' (Late Ordovician/ Early Silurian) Unconformity with Eureka	Very dark dolomite. **Shallow Marine.** conodonts, brachiopods, and corals	Last Chance, Panamint, Argus, Nopah, and Resting Spring Ranges; Cottonwood, Inyo, and Grapevine Mtns.
Rapid sea level rise and major mass extinction Late Ordovician.	*Johnson Spring Fm.-225' (Late Ordovician)	White to light-gray quartzite, minor carbonate, impure quartzite, and siltstone. **Shallow Marine.** Predominately corals.	Inyo Mtns, Waucoba Wash.
	*Barrel Spring Fm.-150'	Dark-gray micaceous hornfels. **Shallow Marine.**	Inyo Mtns, Waucoba Wash.
DV region above sea during mid-Ordovician.	*Eureka Quartzite-350' (Middle Ordovician)	Upper part: white cross-bedded vitreous quartzite, fractures sparkling white. Lower part: hematitic (iron) platy quartzite, sandy and silty dolomite and limestone; No fossils. **Terrestrial/ Marine.**	Last Chance, Argus, Nopah, Panamint, and Resting Spring Ranges; Grapevine Mtns.
Glaciers at south pole (470-255 MY). First land plants. DV on west coast continent of Laurentia..	**Pogonip Group-1500' {Antelope Vly, Ninemile, Goodwin Fms.} (Middle to Early Ordovician)	Minor buff weathering shale beds; crinkled and wavy chert beds; gray sandy and cherty limestone and dolomite. Trilobites, brachiopods, gastropods, nautoloids, and graptolites. **Shallow Marine.**	Last Chance, Panamint, Argus, Nopah, and Resting Spring Ranges; Grapevine Inyo, & Montgomery Mtns.
	**Badger Flat Limestone-500' (Middle Ordovician)	Blue-gray limestone with irregular brown-weathering with silty lenses. **Shallow Marine.**	Inyo Mtns, Waucoba Wash.
	**Al Rose Fm.-375' (Early Ordovician)	Gray, olive, and brown hornsfelsic siltstone and shale, includes carbonate beds. **Shallow Marine.**	Inyo Mtns, Waucoba Wash.
CAMBRIAN PERIOD (542-488 MY)	*Nopah Fm.-1300' (Late Cambrian)	Interbedded light- and dark-gray dolomite, chert nodules, brown siltstone, shale, and shaly and cherty limestone. Brachiopods, trilobites, conodonts, stromatolites in shale. **Shallow Marine.**	Grapevine, Funeral, and Inyo Mtns.; Panamint, Argus, Nopah, Last Chance, Resting Spring Ranges.
Expansion of life in the seas. First marine animals with internal or external skeletons. Predation and burrowing expanded in the seas.	*Tamarack Canyon Dolomite-900'	Light to medium-gray, laminated to massive dolomite, irregular nodules and thin beds of chert. **Shallow Marine.**	Inyo Mtns, Waucoba Wash.
	*Lead Gulch Fm.-320' (Late Cambrian)	Medium gray, flaggy shale; white to gray, coarsely crystalline marble with minor siltstone; medium-gray, thin–bedded. **Shallow Marine.**	Inyo Mtns, Waucoba Wash.
	Bonanza King Fm.-3600' (Middle Cambrian) Unconformity with Nopah.	Broad bands and narrow stripes of dark and light colored limestone and dolomite; shale. Marine environment, bioturbation and trilobites. **Shallow Marine.**	Grapevine, Funeral, Inyo Mtns; Argus, Last Chance, Panamint, Nopah, Kingston, Resting Spring Ranges.
	**Carrara Fm.-1000' (Early Cambrian) Unconformity with Zabriskie	Upper 60% is grayish limestone/dolomite overlying bottom 40% limestone transitioning to siltstone. Cherts from tidal flats. Trans/ Regress marine. Fossils include: stromatolites, oncolites, archaeocyathids, oolites and trilobites. **Shallow Marine.**	Grapevine and Funeral Mtns; Panamint, Nopah, Resting Spring, Kingston, Argus, Last Chance Ranges.
No animal life on land.	**Momola Fm.-1250' (Early Cambrian)	Gray to dark gray shale, siltstone, and buff limestone; gray to blue limestone; gray limestone with inter-bedded gray siltstone and shale. **Shallow Marine.**	Inyo Mtns, Saline Valley.
Burgess Shale deposited (~500 MY).	**Mule Spring Fm.-900' (Early Cambrian)	Blue-gray limestone metamorphosed to marble near plutonic rocks. Trilobites. **Shallow Marine.**	Inyo Mtns, Saline Valley. N. Last Chance Range.
	** Saline Valley Fm.-850' (Early Cambrian)	Gray-black siltstone and quartzite; brown quartzite; brown siltstone and shale. **Shallow Marine.**	Inyo Mtns, Saline Valley. N. Last Chance Range.
	*Zabriskie Quartzite-150' (Early Cambrian) Unconformity Carrara and Wood Cyn	Light gray, thin-bedded to massive quartzite that weathers salmon pink to rusty brown. Vertical wormholes near base, nearshore and farshore fossils. **Shallow Marine.**	Grapevine and Funeral Mtns, Panamint, Nopah, Last Chance, and Resting Spring Ranges.
	*Harkless Fm.-1,500-2000' (Early Cambrian)	Gray-black siltstone and quartzite; brown quartzite; gray-green siltstone and shale. Trilobites. **Shallow Marine.**	Inyo Mtns, Saline Valley. N. Last Chance Range.
	(Uppr) Wood Canyon Fm.-1500' (Early Cambrian)	Sandstone and siltstone with ubiquitous cross bedding. Worm tracks; small, free swimming, marine gastropods; trilobites; brachiopods; echinoderms, and spiral molds. Delta. Tidal. **Shallow Marine.**	See Wood Canyon Fm. (Mid) See Below.
	**Poleta Fm.-1100' (Early Cambrian)	Gray to black quartzitic sandstone; gray shale and inter-bedded siltstone and sandstone. Trilobites, brachiopods. **Shallow Marine.**	Inyo Mountains and Saline Valley. Last Chance Range.
	**Campito Fm.-3500' (Early Cambrian)	Black-quartz mica schist/plutonic rocks; gray shale and inter-bedded quartzitic sandstone and siltstone, gray to black sandstone and inter-bedded gray siltstone and shale. Trilobites. **Shallow Marine.**	Inyo Mtns, Eureka and Saline Valleys. N Last Chance Range.
	**(Mid) Wood Canyon Fm-(Early Cambrian)	Arkosic conglomerate with the final three quarters grading finer into numerous cycles of colorful quartzitic, feldspathic, and arkosic sandstones and maroon, red, purple, etc. siltstones. Cross bedding throughout member. Worm burrows in the upper portion of the middle member. **Primarily Shallow Marine. Brief Terrestrial.**	Alexander and Salt Spring Hills; Grapevine, Owlshead, Funeral Mtns; Panamint, Resting Spring, Last Chance Kingston, Nopah, Ranges.

*** or ** with common shading denotes ~ contemporary formations in different locales. (Not in sequence as table appears to indicate.)**

PROTEROZOIC EON (542-2500 MY)

AGE	FORMATIONS	DESCRIPTION	LOCATIONS
Upper (Late) (800-542 MY) Ediacara-590-635 MY. DV beneath shallow transgressing sea.	**Wood Canyon Fm.- (Lower) 1400'	Siltstone, sandstone, and dolomite with symmetric ripple marks, mudcracks, cross bedding, and possible worm burrows in the upper portion. **Shallow Marine.**	Last Chance Range; Grapevine, Avawatz, and Funeral Mtns; Panamint, Nopah, and Resting Spring Ranges.
	**Deep Spring Fm.- 400'	Dolomite and limestone underlain by gray to black quartzitic sandstone and siltstone. **Shallow Marine.**	Saline and Eureka Valleys; N Inyo Mtns. (Waucoba Rd)
	**Reed Dolomite- 1,400'	Gray to buff oolitic limestone; gray to brown quartzite, buff sandy dolomite, calcerous sandstone; light-gray to cream dolomite. **Shallow Marine.**	Saline and Eureka Valleys; N Inyo Mtns. (Waucoba Rd).
	**Wyman Fm.- 9000'	Gray-blue oolitic limestone to gray limestone; brown to dark gray argillite, brown quartzitic sandstone, gray brown sandstone. **Shallow Marine.**	Eureka Valley; N. Inyo Mtns. (Waucoba Rd).
	Lotus Canyon Fm.-	Limestone and dolomite limestone. **Shallow Marine.**	S. Panamint Range (Manly Peak).
	Stirling Quartzite- 2000' Unconformity with Wood Canyon	Schist; limestone; dolomite; siltstone; quartzite, sandstone. **Primarily fluvial, intermittent marine.**	Grapevine, Funeral, and Avawatz Mtns; Panamint, Kingston, and Nopah Ranges.
	Johnnie Fm.- 4000'	Olive brown and purple shale; tan dolomite; schist, quartzite. Oolites. **Shallow Marine.**	S. Panamint, Resting Spring Kingston, and Nopah Ranges; Avawatz Mtns.
	Ibex Fm.- 1 000'	Conglomerate; limestone; dolomite, shale. **Shallow Marine.**	Owlshead Mtns, Ibex Hills.
	Noonday Dolomite- 1000'	Bedded dolomite; clastic limestone, arkosic sandstone; siltstone. Stromatolites. Ripple marks. **Shallow Marine.** ← **Unconformity between Noonday and Kingston Peak.**	S. Black and Avawatz Mtns; Alexander, Salt Spring, N. Sperry, Ibex, Saddle Peak, and Dublin Hills; Resting Spring, S. Panamint and Nopah Ranges.
Middle (1600-1000 MY) Eukaryotes-1.4 MY 1.1 BY maximum packing of continent Rodinia.. DV beneath sea when rifted ~1.1 BY. Crystal Spring Fm. cherty dolomite was metamorphosed into talc by upwelling diabase.	Kingston Peak Fm.- 3000' to 10,000'(Kingston Range)	Diamictite. Conglomerate, greywacke, limestone. Sandstone, and shale. **Shallow Marine.**	S. Black, Avawatz Mtns; Ibex, Silurian, Saddle Peak, Salt Spring and Valjean Hills; Kingston and Ranges.
	Beck Spring Dolomite- 1000'	Gray dolomite, sandy dolomite. Stromatolites. Pisoliths. **Shallow Marine.**	Saratoga , Saddle Peak, Alexander Hills;Black Mtns; Kingston Range.
	Marvel Limestone	Same as Beck Spring Dolomite except it is located in the Panamints.	Panamint Range.
	Crystal Spring Fm.- 3000' 1.1 BY dated diabase.	Conglomerate, sandstone, dolomite, mudstone, chert, diabase, and talc. Stromatolites. **Marine and terrestrial.** ← **~400 MY Unconformity between Basement and Crystal Spring.**	Black, Avawatz, and Funeral Mtns.; Kingston, Panamint Ranges; Alexander Hills.
Lower (Early) (2500-1600) Prokaryotes: 3.5-3.75 BY photosynthesize and create oxygen. Max Packing of continent Columbia 1.6 BY.	Crystalline Basement (Major constituent of crust of western North America)	Gneiss and schist. Metasedimentary rocks with granitic intrusions. **2.5 to 1.7 BY** "partly" under shallow sea. **Marine and terrestrial.** Extrapolating to **2.5 BY: Terrestrial environment.**	Panamint, Nopah, and Kingston Ranges; Black, Avawatz, and Funeral Mtns.; Alexander Hills.

There are no known rocks older than 2.5 BY in the Death Valley region.

Stratigraphic Column References Include:

Hall, Clarence — "Introduction to the Geology of Southern California and Its Native Plants," 2007.

Jennings, Charles W.; Burnett, John L.; and Troxel, Bennie W. — "Geologic Map of California, Trona Sheet, California Division of Mines and Geology," 1962.

Jennings, Charles W.; — "Geologic Map of California, Kingman Sheet, California Division of Mines and Geology," 1961.

Knott, J. R.; Sarna-Wojcicki, A. et al — "Upper Neogene Stratigraphy and Tectonics of Death Valley" - a review published in: "Fifty Years of Death Valley Research," Elsevier, 2006.

McAllister, James F. — "Geology of the Furnace Creek Borate Area, Death Valley, Inyo County, California, Map Sheet 14," Calif. Div. of Mines, 2nd 1977.

Reynolds, Mitchell — "Stratigraphy and Structural Geology of the Titus and Titanothere Canyons Area, Death Valley California," Dissertation Doctor of Philosophy, University of California Berkely, 1969.

Sarna-Wojcicki, Andrei et al — "Tephra Layers of Blind Spring Valley and Related Upper Pliocene and Pleistocene Tephra Layers, CA, UT, NV: Isotropic Ages, Correlation, and Magnetostratigraphy," *USGS PP 1701 2005.*

Stran, Rudolph G. — "Geologic Map of California, Mariposa Sheet, California Division of Mines and Geology," 1967.

Strietz, Robert and Stinson, Melvin C. — "Geologic Map of California, Death Valley Sheet, California Division of Mines and Geology," 1977.

Troxel, Bennie W. — Verbal discussions, August 2009.

Walker, J.D.; Geissman,J.W. — 2009 Geologic Time Scale, GSA, 2009.

References

Blakey, Ronald C.	Paleogeography of the Southwestern United States website, Northern Arizona University.
Calzia, James Peter	"Geologic Studies in the Kingston Range, Southeastern Death Valley;" Doctoral Thesis; University of California, Davis; 1997.
Calzia, James Peter; Editor	"Fifty Years of Death Valley Research," Elsevier, 2006.
Diehl, Paul	"Stratigraphy and Sedimentology of the Wood Canyon Formation, Death Valley," California, 70th Annual Meeting of the Cordilleran Section of the Geological Society of America, Field Trip 1, 1974.
Fairchild, James	Harmony Borax Works gathering, processing and shipping, written correspondence, emails July, 2009.
Fiero, Bill	"Geology of the Great Basin," University of Nevada Press, 1986.
Garrett, Donald E.	"Borates," Academic Press, 1998.
Gomez, F.; Hsieh, J.; Holt, B.; Murray B.; Kirschvink, J.	"Outcrop Geology of Plio-Pleistocene Strata of the Confidence Hills, Southern Death Valley," California; Division of Geological and Planetary Sciences, California Institute of Technology.
Jennings, Charles W.; Burnett, John L.; and Troxel, Bennie W.	"Geologic Map of California, Trona Sheet," California Division of Mines and Geology, 1962.
Stran, Rudolph G.	"Geologic Map of California, Mariposa Sheet," California Division of Mines and Geology, 1967.
Strietz, Robert and Stinson, Melvin C.	"Geologic Map of California, Death Valley Sheet," California Division of Mines and Geology, 1977.
Jennings, Charles W.;	"Geologic Map of California, Kingman Sheet," California Division of Mines and Geology, 1961.
Kleinhampl, Frank J. and Ziony, Joseph I.	"Geology of Northern Nye County, Nevada," Nevada Bureau of Mines and Geology, Bulletin 99A.
Knott, J. R.; Sarna-Wojcicki, A. M.; Machette, M. N.; Klinger, R. E.	"Upper Neogene stratigraphy and tectonics of Death Valley - a review," published in: "Fifty Years of Death Valley Research," Elsevier, 2006.
McAllister, James F.	"Geology of the Furnace Creek Borate Area, Death Valley, Inyo County, California, Map Sheet 14," California Division of Mines, Second Printing 1977.
Miller, Julia M. G.	"Stratigraphy and Sedimentology of the Upper Proterozoic Kingston Peak Formation, Panamint Range, Eastern California," Doctoral Thesis; July, 1983.
Morrison, Roger B.	"Lake Tecopa: Quaternary geology of Tecopa Valley, CA, a multimillion year record and its relation to the proposed nuclear-waste repository at Yucca Mountain, Nevada," GSA Special Paper 333, 1999.
Reynolds, Mitchell	"Stratigraphy and Structural Geology of the Titus and Titanothere Canyons Area, Death Valley California," Dissertation Doctor of Philosophy, University of California Berkely, 1969.
Roberts, Michael T.	"Stratigraphy and Depositional Environments of the Crystal Spring Formation, Southern Death Valley Region, California," 70th Annual Meeting , Cordilleran Section of the Geological Society of America, Field Trip 1, 1974.
Rogers, J. J. W.; Santosh, M.	"Continents and Supercontinents," Oxford University Press, 2004.
Sarna-Wojcicki, Andrei et al	"Tephra Layers of Blind Spring Valley and Related Upper Pliocene and Pleistocene Tephra Layers, CA, UT, NV: Isotropic Ages, Correlation, and Magnetostratigraphy." *USGS PP 1701 2005.*
Scotesse Christopher	Paleomap Project, Earth History, http://www.scotese.com (Paleomap website).
Walker, J.D.; Geissman, J.W.	2009 Geologic Time Scale, GSA, 2009.
Wertz, William Earl	"The Depositional Environments and Petrography of the Stirling Quartzite, Death Valley Region, California and Nevada," Pennsylvania State University doctoral thesis, June 1983.
Williams, Eugene G.; Wright, Lauren A.; Troxel, Bennie W.	"The Noonday Dolomite and Equivalent Stratigraphic Units, Southern Death Valley Region, California," 70th Annual Meeting of the Cordilleran Section of the Geological Society of America, Field Trip 1, 1974.
Wright, Lauren A.; Troxel, Bennie W.	"Geology of the North 1/2 Confidence Hills 15' Quadrangle. Inyo County, California, California Map Sheet 34," California Division of Mines and Geology, 1984.
Wright, Lauren A.; Troxel, Bennie W.; Williams, Eugene G.; Roberts, Michael T.; Diehl, Paul E. ;	"Precambrian Sedimentary Environments of the Death Valley Region, Eastern California," 70th Annual Meeting of the Cordilleran Section of the Geological Society of America, Field Trip 1, 1974.
Wright, Lauren A.; Williams, Eugene G.; Cloud, Preston	"Stratigraphic Cross Section of Proterozoic Noonday Dolomite, War Eagle Mine Area, Southern Nopah Range, Eastern California," 70th Annual Meeting of the Cordilleran Section of the Geological Society of America, Field Trip 1, 1974.
Wright, Lauren A.;	"Talc Deposits of the Southern Death Valley Kingston Range, Region, California," Special Report 95, California Division of Mines and Geology, 1968.
Wright, Lauren A. ; Troxel, Bennie W.	"Geology of Southern California, Bulletin 170, Geologic Guide No. 1, Western Mojave Desert and Death Valley Region," California Division of Mines, 1954.
Wrucke, Chester T.; Stone, Paul; Stevens, Calvin H.	"Geologic Map of the Warm Spring Canyon Area, Death Valley National Park, Inyo County, California, Map 2974 Version 1.1," USGS, 2008 Revision.

Glossary

Terms	Definitions
andesite	Dark-colored, fine grained, extrusive, igneous rock. When porphyritic contains phenocrysts, primarily composed of plagioclase and one or more of the mafic minerals (e.g. biotite, hornblende, pyroxene) with a matrix of the same composition as the phenocryst. The extrusive equivalent of diorite.
archaeocyathid	Marine organism characterized by a cone-, goblet- or vase-shaped skeleton. World wide distribution in the Cambrian.
ash fall	A rain of airborne volcanic ash falling from an eruption cloud.
arkosic	Primarily quartz. feldspar rich (>25%).
ash flow	Generally a highly heated mixture of volcanic gases, ash, crystals, and volcanic rock fragments that travel down the flanks of a volcano or along the ground.
basalt	Dark colored, extrusive, igneous rock composed primarily of calcic plagioclase and pyroxene.
biotite	A black, common rock forming mineral of the mica group. Perfect basal cleavage.
bioturbation	The turning and churning of sediments by a burrowing organism.
borate	A mineral compound containing boron and characterized by BO_3^{-3}. 238 known borates.
boulder	A detached rock mass larger than a cobble, having a diameter 256 mm (10 inches) or about the size of volleyball, being rounded or otherwise distinctly shaped by abrasion in the course of transport.
brachiopod	Bivalve (double-shelled) marine invertebrates. Opposing shells are not mirror images as in petycypods (such as modern clams). Common and widespread in the Paleozoic but fewer exist today.
braided stream	A stream that divides into an interlacing network of branching and reuniting shallow channels separated from each other by islands or channel bars, resembling in plan strands of a complex braid.
breccia	A coarse grained sedimentary rock with angular rock fragments.
BY	Billion years or billion years ago.
calcite	$CaCO_3$. Calcium carbonate. A common rock-forming mineral that is a chief constituent of limestone, tuffa, travertine.
caldera	A large basin-shaped volcanic depression, more or less circular.
caliche	Soil, gravel, rocks cemented by $CaCO_4$.
carbonate	Sediment formed of the carbonates (CO_3) of calcium, magnesium, and/or iron, e.g. limestone.
clast	An individual fragment within a sedimentary rock.
Columbia	Ancient supercontinent. Max. Packing ~1.6 BY.
creodont	Extinct carnivorous mammal. ~55-35 MY.
cross-bedded	Arrangement of strata inclined at an angle to the main stratification.
cyanobacteria	A bacteria that derives its energy from photosynthesis.
clastic	Sedimentary rock comprised principally of fragments derived from pre-existing rocks and transported mechanically to their places of deposition (e.g. sandstone, shale, or conglomerate or a limestone).
cobble	A rock fragment between 64 and 256 mm (2 1/2 to 10-inches) in diameter thus larger than a pebble and smaller than a boulder, rounded or otherwise abraded in the course of aeolian, aqueous, or glacial transport.
conglomerate	A coarse grained sedimentary rock with rounded rock fragments of granule or larger size.
crystal	A homogeneous, solid body of a chemical element, compound, or isomorphous mixture, having a regularly repeating atomic arrangement that is outwardly expressed by plane faces.
dacite	A fine-grained extrusive rock with the same general composition as andesite but having a less calcic plagioclase and more quartz.
diabase	An intrusive rock consisting essentially of a feldspar mineral of the plagioclase series having approximately equal proportions of sodium and calcium.
diamictite	Coarse, angular to well-rounded clastic fragments supported in a fine matrix.
dolomite	A sedimentary rock formed from limestone when part of the calcium in the limestone is replaced by magnesium.
echinoderm	Any solitary marine bottom dweller invertebrate, belonging to the phylum Echinodermata, characterized by radial symmetry, an endoskeleton formed of plates or ossicles of crystalline calcite, and a water vascular system. Crinoids belong in this phylum.
ediacara	The first true multi-cellular animals which appeared at the end of the Proterozoic, ~635-590 MY.
epidote	Green metamorphic rock derived from limestone.
erosion	The removal and transportation of rocks by water, winds, and/or ice.
eukaryote	A type of living cell containing a true nucleus enclosed within a nuclear membrane, and having well-defined chromosomes and cell organelles. Life form, which appeared about ~1.8 By.
extension	Refers herein to the stretching of the earth's crust. In the Great Basin region it is approximately in an east-west direction and results in the basins (e.g. Death Valley) and ranges (e.g. Panamint Range) found throughout the Great Basin.
facies	The aspect, appearance, and characteristics of a rock unit, usually reflecting the conditions of its origin.
fanglomerate	A sedimentary rock of heterogeneous materials that were originally deposited in an alluvial fan and have since become cemented into solid rock.
fault	Break in the earth's crust across which rocks are dis-placed. {Normal=Tension, Reverse= Compression}.

feldspar	A group of abundant rock-forming minerals of the general formula $Mal(Al,Si)_3O_8$ where M can be K, Na, Ca, Ba, Rb, Sr, or Fe. Constitutes 60% of Earth's crust.
flaggy	Splits uniformly along bedding planes into thin slabs that may be suitable for terraces and walls.
fluvial	Of or pertaining to rivers, produced by the action of a stream or river.
fold	A bend in the earth's crust (e.g. anticline, syncline).
foliation	The planar structure that results from the flattening of the constituent grains oif a metamorphic rock.
Fm., formation	A formally defined body of rock strata that consists dominantly of a certain lithologic type or combination of types, or has other unifying lithologic features. Some formations can be subdivided into "members".
gabbro	A group of dark colored basic igneous rocks that are the intrusive counterpart of basalt.
gastropods	Commonly known as snails or slugs. Marine, freshwater, terrestrial varieties.
geology	The study of Planet Earth, the materials of which it is made, the processes that act on those materials, the products formed, and the history of the planet and its life forms since its origin.
gneiss	A foliated rock formed by regional metamorphism. Commonly feldspar and quartz rich.
granite	A coarse grained, plutonic rock in which quartz makes up 10-50% of the light colored minerals, which are also composed mostly of potassium and sodium rich feldspars.
hornblende	Black. Very common in gabbro. $A_2B_5(Si,Al)_8O_{22}(OH)_2$ where A is Mg, Fe, Ca, or Na and B is mainly Mg, Fe_3^{+2}, Al, and Fe^{+3}.
halite	"Table" salt. Sodium chloride. NaCl.
igneous rock	One of the three main classes into which rocks are divided, the others being sedimentary and metamorphic. Igneous rock has solidified from molten or partly molten materials (magma or lava). Intrusive (plutonic) igneous rocks cool deep within the earth's crust and contain very visible crystals (e.g. granite, diorite, or gabbro)). Extrusive (volcanic) rocks cool at or above the surface producing very fine crystals, glass, or a mixture of volcanic rock fragments and ash (e.g. obsidian, rhyolite, andesite, or basalt).
interfinger	To grade or pass from one material into another through a series of interpenetrating wedge-shaped layers. Looks like fingers on two hands interlaced.
intrusive	Plutonic igneous rocks. Cool deep within the earth's crust and contain very visible crystals (e.g. granite, diorite, or gabbro).
KY	Thousand years or thousand years ago.
lacustrine	Pertaining to, produced by, or inhabiting a lake or lakes.
latite	An extrusive igneous rock (volcanic) that has large crystals of plagioclase $((Na,Ca)Al(Si,Al)Si_2O_8$ and orthoclase $(KalSi_3O_8)$ in equal amounts, little quartz, commonly hornblende and scattered pyroxene in a finely crystalline matrix.
lava	Molten rock that issues from a volcano or a fissure on the earth's surface.
Liesegang rings	Secondary, nested rings or bands caused by rhythmic precipitation within a fluid saturated rock.
limestone	A sedimentary rock composed mostly of calcite $(CaCO_3)$, precipitated in shallow seawater-through the actions of organisms or comprised of fragments from calcite secreting organisms or processes.
lithologic	Pertaining to rocks. Physical characteristics such as color, mineral composition, and grains.
lithification	The conversion of newly deposited sediments or volcanic deposits into a rock.
magma	Naturally occurring molten rock material, generated within the earth.
massive	Said of rocks of any origin that are more or less homogeneous in texture or fabric and do not split easily. They display an absence of flow layering, foliation, cleavage, joints, fissility, or thin bedding.
member	A lithostratigraphic unit of subordinate rank, comprising some specially developed part of a formation. It may or may not be formally defined or mappapable.
metavolcanic	Volcanic rocks that show evidence of having been metamorphosed.
micaceous	Consisting of or resembling mica. Capability of easily being split into thin sheets.
multituberculates	Extinct, rodent-like mammals. Not placental but egg laying. ~160-35 MY.
metamorphism, metamorphic rock	A rock having gone the transformation from a preexisting rock into a texturally or minerallogically distinct new rock as a result of high temperature, high pressure, or both but without melting in the process (e.g. marble, slate, quartzite, gneiss, schist), generally forms at depth in the earth's crust.
mine	A site, open pit or vertical or horizontal excavation into the earth, where mineral resources are extracted for the purpose of financial or other gain.
mineral	A naturally occurring (not man made) inorganic element or compound having an orderly internal structure and characteristic chemical composition, crystal form, and physical properties (e.g. borax, salt, limestone).
mudstone	A sedimentary rock hardened by pressure and comprised of equal proportions of clay and silt and lacking the characteristic of splitting into thin, flat plates like shale.
MY	Million years or million years ago.
oncolite	Formed by accretion of successive layered masses of gelatinous blue-green algae. Smaller than a stromatolite. Less than 10 cm in diameter.
oolite	Small, rounded carbonate rock particles consisting of a nucleus (rock fragment) and concentrically layered calcium carbonate formed by accretion in calcium carbonate saturated waters of tidal flats. Look like fish eggs with a diameter of .25 to 2.0 mm.
orogeny	The process of formation of mountains.
orthoclase	A white, pink, or gray mineral of the alkali feldspar group: $KAlSi_3O_8$. Common in granites, acid igneous rocks, and crystalline schists.

outcrop Part of a geologic structure or formation that appears at the surface of the Earth.

paleogeography The study of the physical geography of all or part of the Earth's surface at some time in the geologic past.

paleontology The study of life in past geologic time, based on fossil plants and animals and their relationship to current plants and animals.

Pangea A supercontinent that existed from about ~300 to 200 million years ago and included most of the Earth's crust.

passive (continental) margin A margin that includes continental shelf, slope, and rise that generally extends down to an abyssal plain at a depth of 5-kilometers. Occurs where tectonic plates are diverging and not converging. Usually no earthquakes or volcanoes.

pebble A rock fragment, generally rounded by abrasion, larger than a granule and smaller than a cobble; it has a diameter of 4 to 64 mm (5/32-inches to 2 1/2-inches), or a size between that of a pea and a tennis ball.

perlitic A feature of glassy igneous rocks that have cracked due to contraction during cooling, the cracks forming small concentric pearl-like spheroids.

phenocryst One of the relatively large and ordinarily conspicuous crystals of the earliest generation in a porphyritic igneous rock.

pisolith An accretionary body in a sedimentary rock, resembling a pea in size and shape, and constituting one of the grains that make up a "pisolite rock." Often formed of calcium carbonate, and some are thought to have been produced by bio-chemical algal-encrustation process. Larger than an oolith.

plagioclase A group of triclinic feldspars of general formula $(Na,Ca)Al(Si,Al)Si_2O_8$. Among the commonest rock-forming minerals. White in color.

plate tectonics A theory that the earth's surface is divided into large, thick plates that are moving and changing in size by collision, melting, fragmentation, and warping. Intense geologic activities occur at the plate boundaries.

platy Splits uniformly along bedding planes. Thinner than slabby or flaggy.

pluton, plutonic An igneous intrusion (e.g. body of magma rising and forcing its way into existing rocks). Cools below surface. Granite, diorite, gabbro.

pluvial Geological process or feature resulting from rain.

porphyritic An igneous rock of any composition that contains conspicuous phenocrysts in a fine-grained matrix.

pyroxene Dark mineral with silicate, iron, and magnesium.

quartz Crystalline silica (SiO_2). Next to feldspar, it is the commonest mineral.

quartzite A metamorphosed rock, consisting mainly of quartz, that was originally sandstone (e.g. metaquartzite) or sandstone consisting of quartz grains cemented by secondary silica (e.g. orthoquartzite).

receptaculitid There is a debate as to whether this was a sponge or algae. Ordovician to Permian.

rhyolite Extrusive igneous equivalent to intrusive igneous granite. Typically porphyritic and commonly exhibiting flow texture, with phenocrysts of quartz and alkali feldspar in a glassy matrix.

rift A major linear depression in the earth's crust formed by tension pulling apart a section of the earth's crust.

rock An aggregate of one or more minerals or solid organic material.

Rodinia An ancient supercontinent. Tectonic forces assembled and disassembled this continent ~1.1 BY and 750 MY, respectively. North America was once part of Rodinia.

sandstone A sedimentary rock formed by cementation of sand size grains in a matrix of silt or clay and cemented by silica, iron oxide, or calcite. The consolidated equivalent of sand.

savannah An open, essentially treeless, grassy plain. Usually found in tropical or subtropical regions.

sedimentary rock Formed from pre-existing rocks (sandstone) or pieces of once living organisms (limestone). Chemical sediments or rocks form from precipitation of minerals from solution (e.g. calcium carbonate, borax, salt, gypsum). Organic sediments form from the remains of once living organisms (e.g. limestone, coal).

shale Fine grained sedimentary rock formed by the compaction of clay, silt, or mud. It readily splits into very thin, flat layers. Deposited on lake bottoms, at the ends of rivers, in deltas, and on quiet parts of the deep ocean floor.

silicate A compound whose structure contains silicon and oxygen arranged in tetrahedron form, SiO_4. Largest and most common class of minerals (orthoclase and plagioclase that are called feldspars, quartz, hornblende, etc.).

siliceous Containing abundant silica (SiO_2) such as chert, chalcedony, or quartz.

siltstone A sedimentary rock consisting of silt sized grains. It has the same texture and composition of shale but lacks its ability to split into very thin layers. Usually splits flaggy.

Stratigraphic column The sequence of rock strata described or illustrated in a vertical column.

stromatolite Layered structures formed by the trapping of sedimentary particles and precipitation of calcium carbonate in response to the metabolic activities and growth of mat-like colonies of cyanobacteria and some other prokaryotes.

subduction The process of one lithospheric plate descending beneath another (e.g. heavy oceanic plate dives beneath light continental plate).

talc An extremely soft, light green or gray mineral $Mg_3Si_4O_{10}(OH)_2$. Soapy feel. Easily cut with a knife.

Titanotheres Eocene herbivorous mammals that somewhat resembled today's rhinoceros (s). Also known as brontotheres and meaning Sioux for "Thunder Beast." The largest of these animals were 8' tall at the shoulder, 12'-14' long, and weighed 4-5 tons.

tectonic forces Forces generated within the earth that result in uplift, movement, or deformation of part of the earth's crust.

tephra A collective term for all clastic materials ejected from a volcano and transported through the air.

Tertiary A period of time in the Cenozoic era (0-65.5 MY) that is before the Quaternary (0-2.6 MY). It is the period of time from 2.6 to 65.5 MY.

turbidity current A bottom flowing current laden with sediment, moving swiftly down a subaqueous slope and spreading horizontally on the floor of the body of water.

topography The general configuration of a land surface, including its relief and position of its features.

trachyandesite An extrusive igneous (volcanic) rock, intermediate in composition between trachyte and andesite, with sodic plagioclase, alkali feldspar, and one or more mafic minerals (biotite, amphibole, or pyroxene).

trachyte A group of fine-grained, generally porphyritic, extrusive igneous (volcanic) rocks having alkali feldspar and minor mafic minerals as the main components, and possibly a small amount of sodic plagioclase.

travertine A porous deposit of calcite that often forms around hot springs.

trilobite Paleozoic marine organism resembling today's sow bugs. It lived from the lower Cambrian through the Permian. It is characterized by a three-lobed ovoid outer skeleton, divided lengthwise into axial and side regions and transversely into cephalon (head), thorax (middle), and pygidium (tail).

tuff General term for all consolidated pyroclastic rocks having glass shards as a major constituent.

unconformity A break or gap in the geologic record.

vitrophyre Any porphyritic igneous rock having a glassy matrix.

volcanic Pertaining to the activities, structures, or rock types of a volcano.

weathering The physical disintegration and chemical decomposition of rock, by exposure to atmospheric agents near the earth's surface, with little or no transport.

welded tuff A pyroclastic rock that has been indurated by the welding together of glass shards, and possible crystals and volcanic rock fragments, under the combined action of the heat retained by particles, the weight of overlying material, and hot gases The rock generally appears banded or streaky.

xenolith Fragment of rock distinct from the igneous rock in which it is enclosed.

A winter rain in the Greenwater Range erodes the mountains and fills the basins.

Index

Index

Index

T

U

The narrows at Hole-in-the-Wall, Funeral Mountains, frames a broad vista with the Paleozoic Nopah Fm. crowning the Bonanza King Fm. and the colorful Zabriskie Quartzite. Photograph to the right illustrates why Hole-in-the-Wall got its name. The tilted beds of an ancient alluvial fan are the coarse fluvial conglomerates of the upper Furnace Creek Fm. The holes result from the weathering out of boulders. At this location, lacustrine mudstone abruptly intertongues with the conglomerate.

Bennie Troxel: "Hope you enjoyed the trip."